French Foreign Legion

MINES AND BOOBY TRAPS

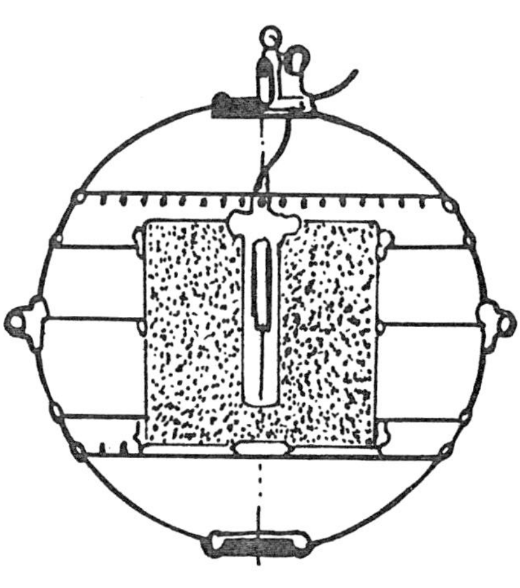

PALADIN PRESS
BOULDER, COLORADO

French Foreign Legion:
Mines and Booby Traps

ISBN 0-87364-344-5
Printed in the United States of America

Published by Paladin Press, a division of
Paladin Enterprises, Inc., P.O. Box 1307,
Boulder, Colorado 80306, USA.
(303) 443-7250

Direct inquiries and/or orders to the above address.

CONTENTS

Surprise or be surprised!

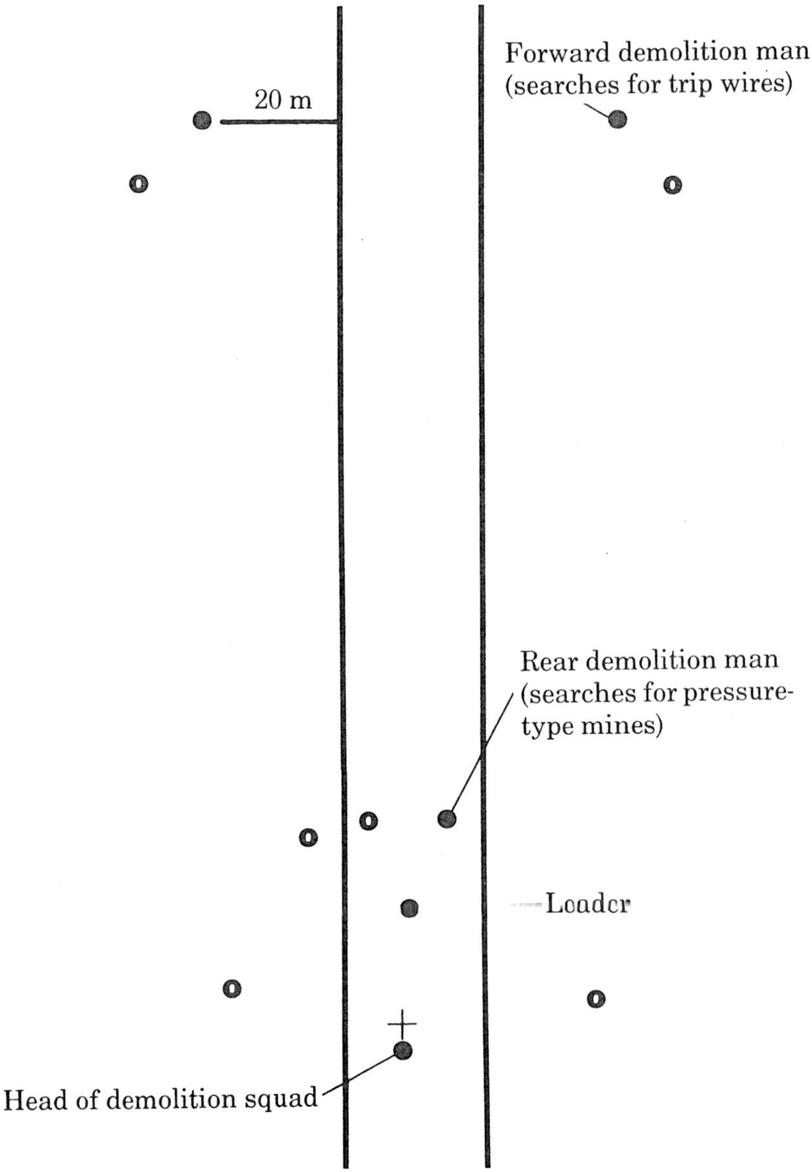

Demolition Squad. Plan of Operation.
● Demolition crew ○ Protection personnel.

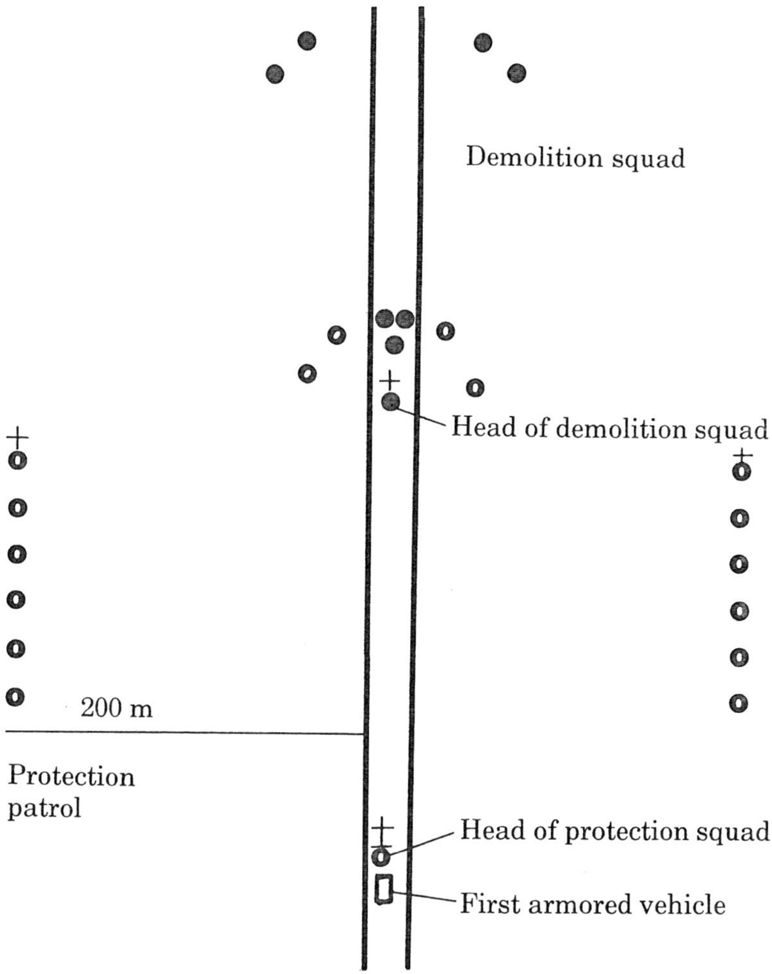

Protection Squad. Plan of Operation.
● Demolition crew ○ Protection personnel.

Beware! The piece of string may be a booby trap.

Beware! If you step on a tiger's tail you may have one chance in a hundred of surviving. If you step on a mine, you won't have one chance in a thousand.

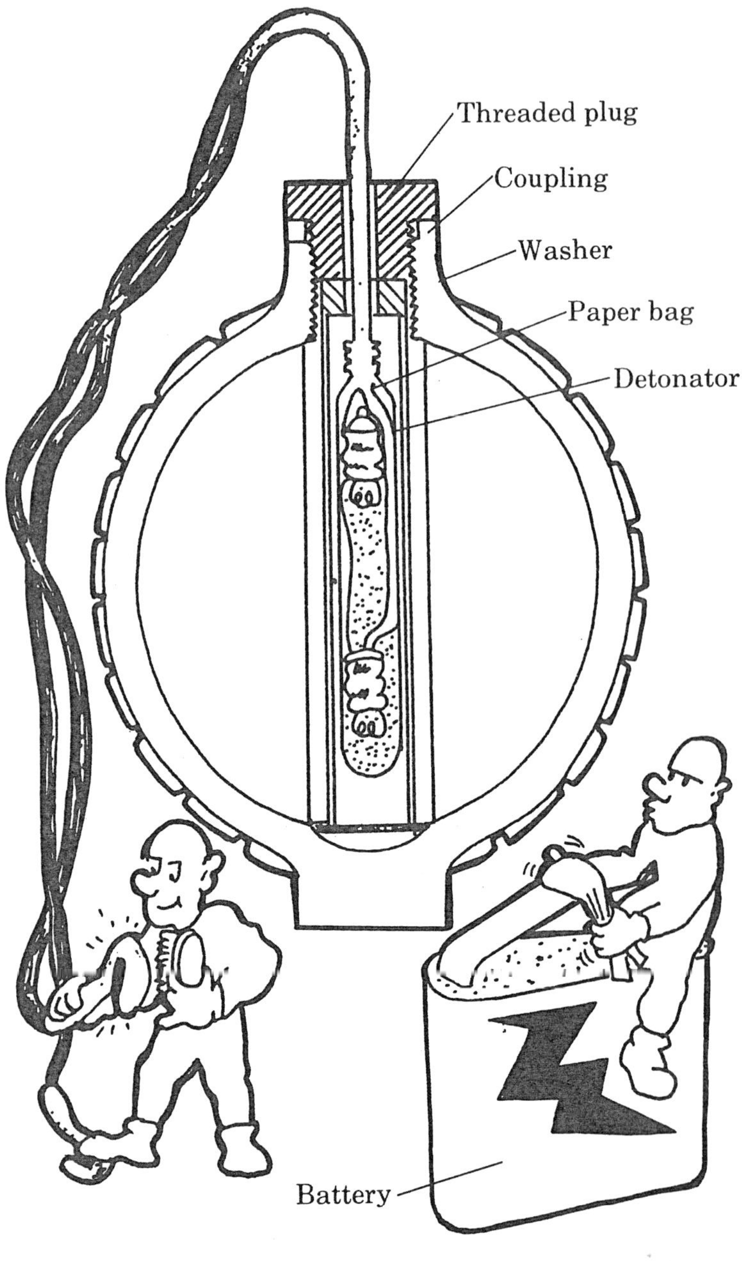

VC Spherical Mine with Electric Fuze. Scale: 1:2.

VC* SPHERICAL MINE WITH ELECTRIC FUZE

Type: Fragmentation antipersonnel mine.

Appearance: Mine is made of cast iron with serrations for fragmentation effect.

Size: Outer diameter = 18 cm.

Firing device: Improvised electric fuze. Double firing with two flashlight bulb caps.

To set: Place the bag containing the improvised electric fuze into the chamber. Screw in the threaded plug.

Functioning: Connecting the conductor wires to the terminals of an electric source of more than 3.5 volts causes the explosion.

Neutralizing: Cut the electric wires one at a time. Unscrew the cap. Remove the bag by rotating the mine. Do not exert any pull on the wires. If the bag does not come out, destroy the mine on the spot.

Source: Company of the Corps of Engineers 71/1, Region of Sadec, March, 1948.

---

*Translator's note: V. M. in the original French, which stands for Viet Minh (North Vietnamese). Present-day meaning would be Viet Cong (VC).

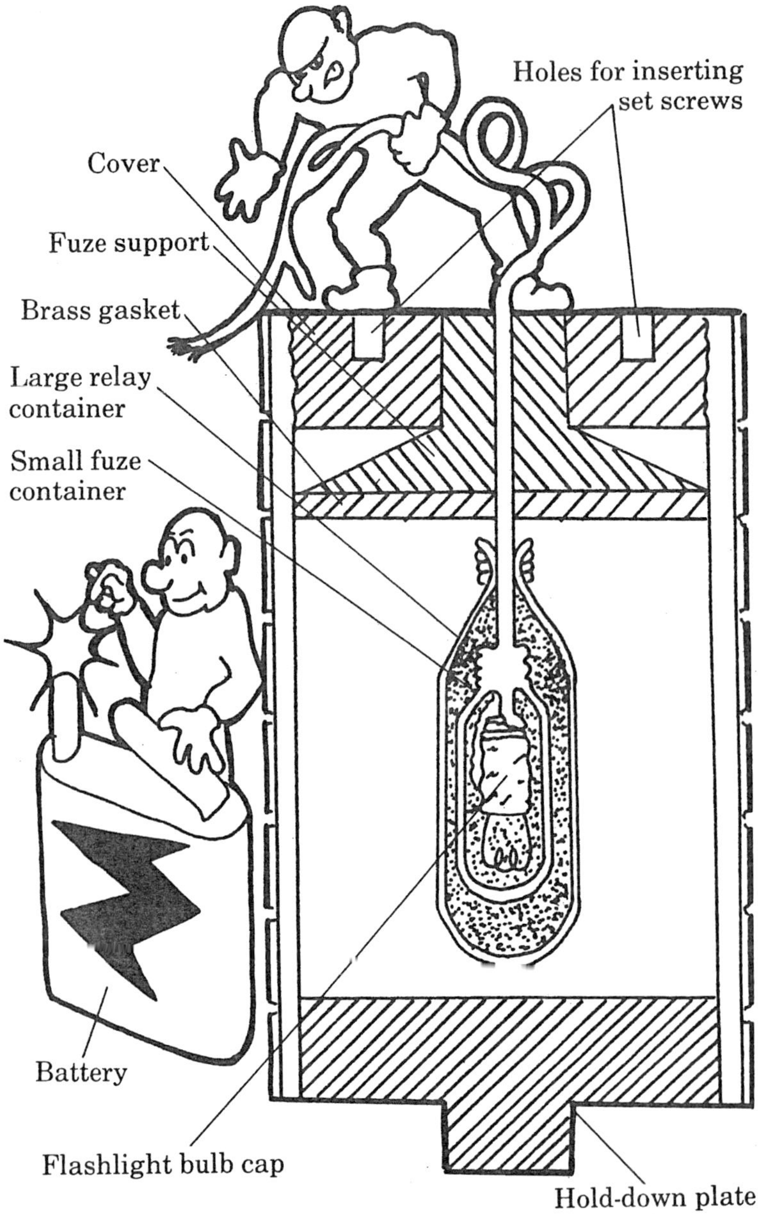

Fragmentation Mine, 80 mm Outer Diameter.

FRAGMENTATION MINE

Type: Electric antipersonnel mine. Remotely detonated by an operator.

Appearance: Cast-iron cylinder with square cross section. Two electric wires come out of the cap.

To set: Delivered ready to operate. Connect the conductor wires to the wires of the mine.

Functioning: Connecting the conductor wires to the terminals of an electric battery greater than 3.5 volts detonates the mine.

Neutralizing: Cut the electric wires one by one. Carry it away, if necessary, and destroy.

Caution: A strong pull on the electric wires can cause an accidental explosion.

Source: T.F.I.S., 1948.

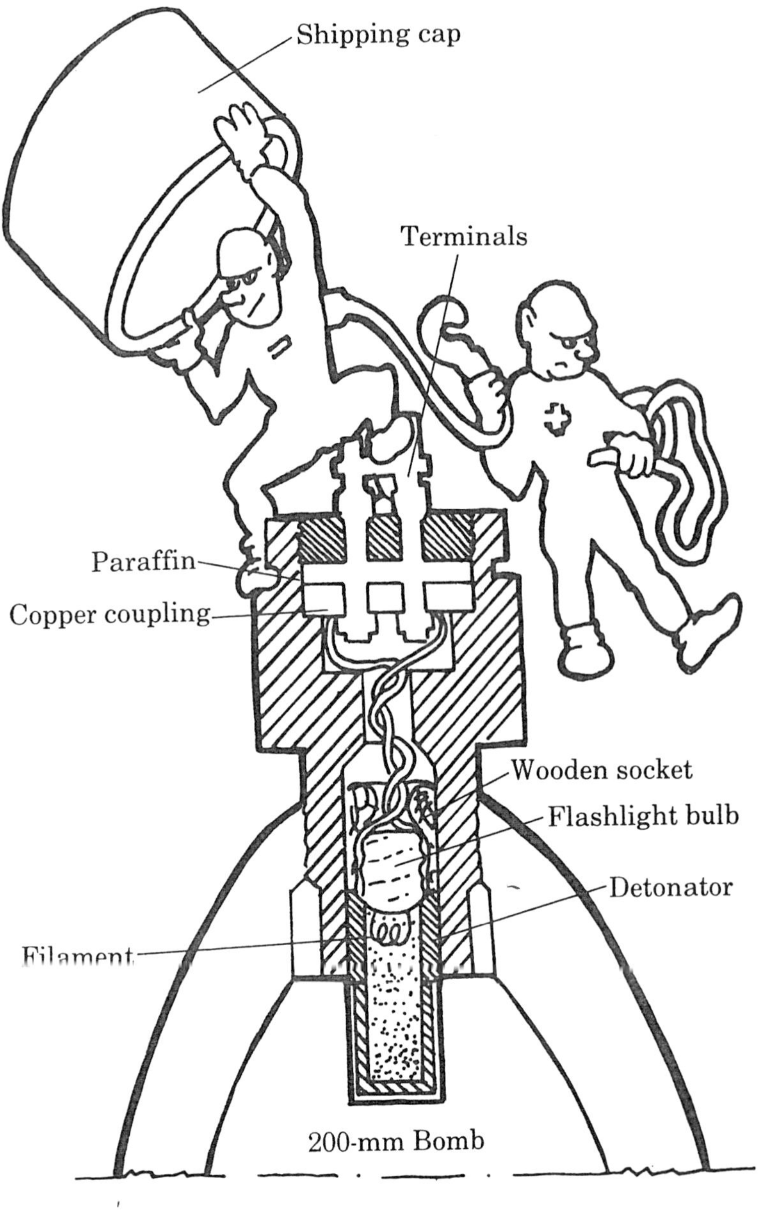

VC Electric Mine Fuze.

VC ELECTRIC MINE FUZE

Type: Improvised electric fuze.

Employment: For antivehicular mine.

Appearance: Made of machined copper with two screw-type terminals on top.

To set: Remove the shipping cap protecting the terminals. Insert into the chamber of a 200-mm bomb. Attach the lead wires to the terminals.

Functioning: Connecting the conductor wires to the terminals of an electric battery greater than 3.5 volts causes the fuze to ignite.

Neutralizing: Cut the wires one by one. Take out the fuze.

Source: T.F.I.S., September, 1948.

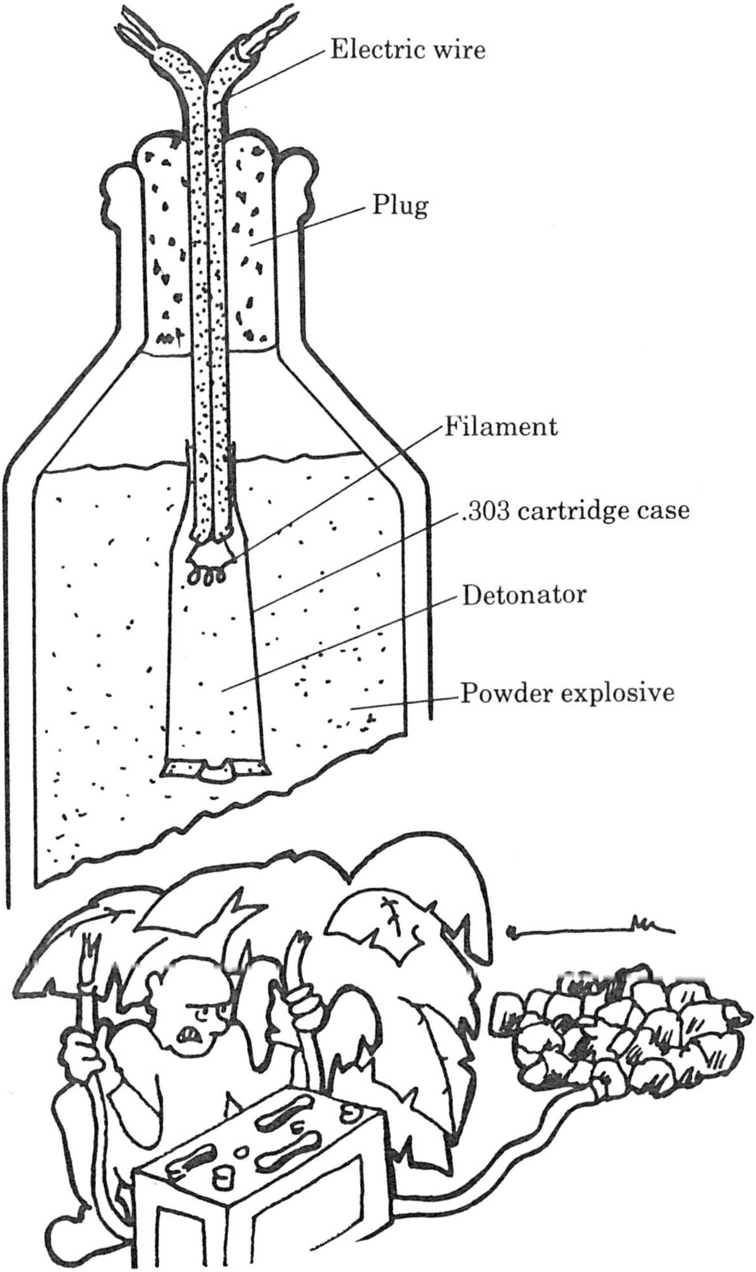

Bottle-Type Mine.

IMPROVISED BOTTLE-TYPE MINE

Type: Improvised electric fuze.

Appearance: Bottle or glass container.

Functioning: Connecting the lead wires to the terminals of an electric battery causes the fuze to ignite.

Neutralizing: Cut the wires one by one. Destroy the mine on the spot, if possible. Otherwise, carry the entire mine away and destroy it. Do not pull on the wires.

Source: 3rd Infantry Regiment, Tonkin. March, 1948.

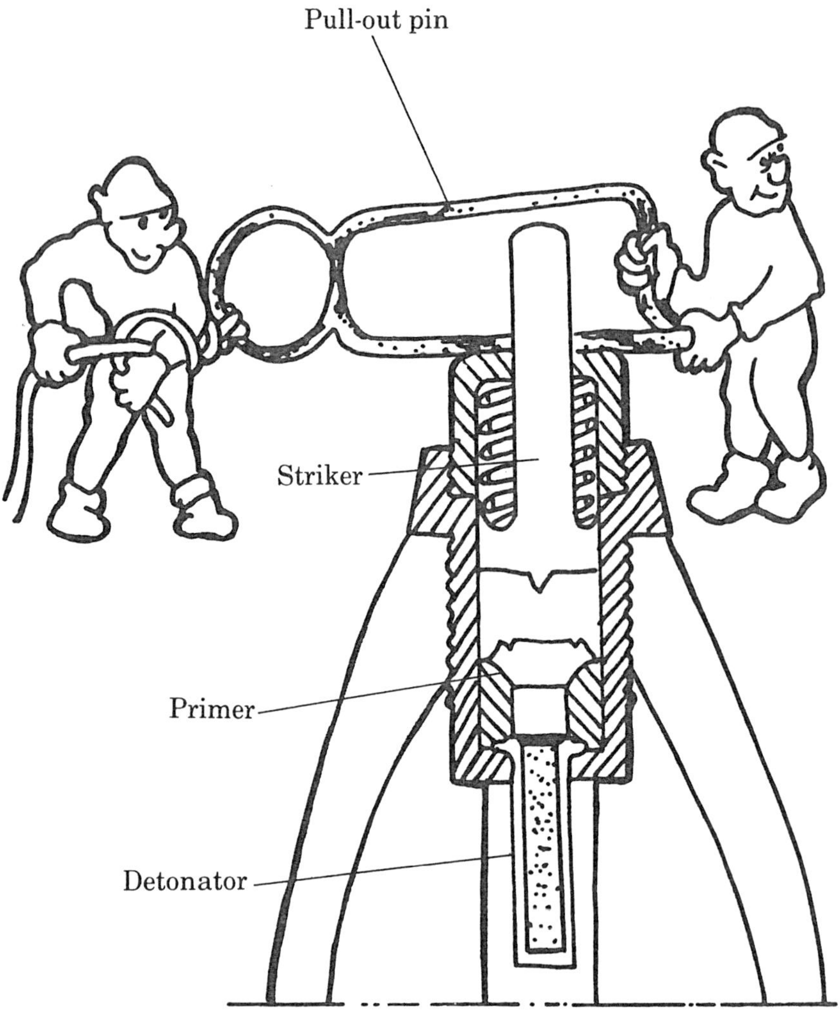

Pull-Type Fuze For 88-mm Shells. Scale: 1:1.

PULL-TYPE FUZE I

Type: Trigger pin.

Appearance: Made of machined copper. The trigger pin draws back a safety pin.

To set: Insert into the chamber of an 88-mm shell. Attach the pull cord. Unfasten the safety part of the pin and twist it toward the ring.

Functioning: In the armed position, the striker spring is compressed. The trigger pin passing through the end of the striker holds it down. A pull on the wire releases the pin. The striker is freed and hits the primer.

Neutralizing: Cut the pull cord. Untwist the safety part of the pin and reset the fastener. Remove the fuze.

Caution: If the pin does not pass entirely through the end of the striker, do not attempt to replace it.

Source: T.F.I.S., 1948.

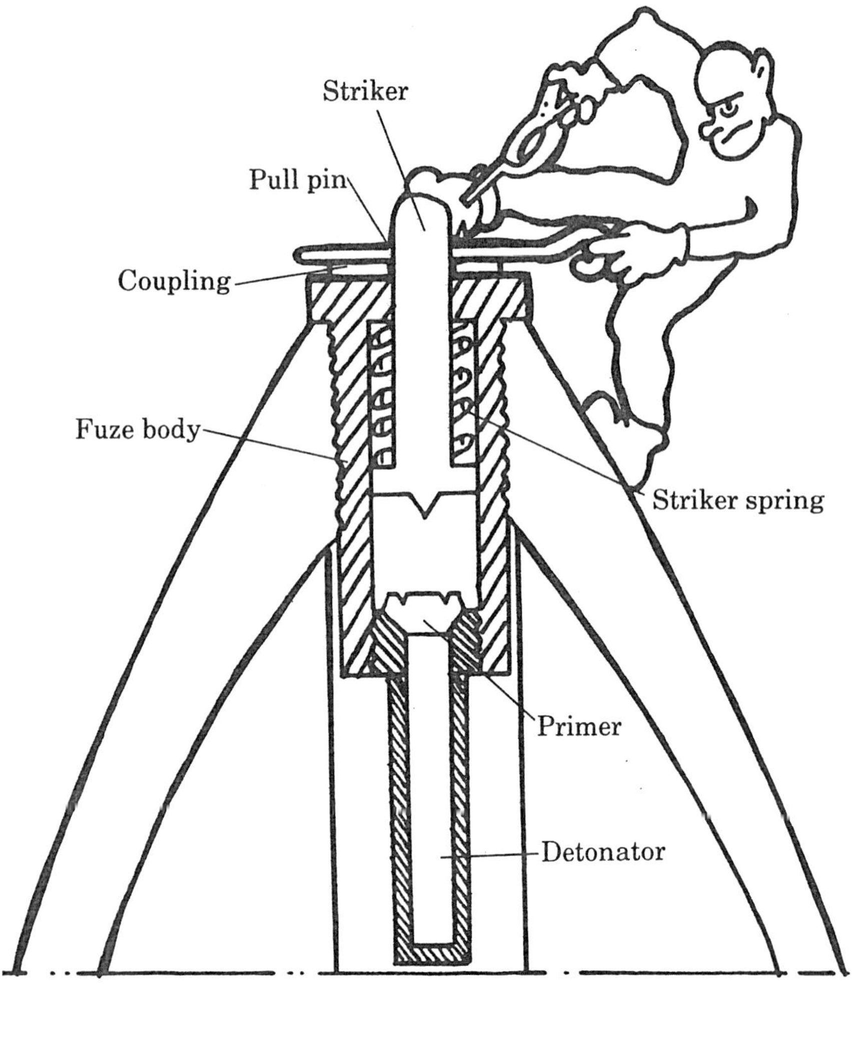

Pull-Type Fuze For 240-mm Shells. Scale: 1:1.

PULL-TYPE FUZE II

Type: Trigger pin.

Appearance: Made of machined copper. The pin is inserted through the visible end of the striker.

To set: Place the fuze inside the chamber of a 240-mm shell. Attach the pull cord to the ring on the end of the pin.

Functioning: In the armed position, the striker spring is compressed. The trigger pin inserted through the end of the striker holds it down. A pull on the cord releases the pin. The striker is freed and hits the primer.

Neutralizing: Cut the pull cord. Bend back the end of the pin.

Caution: If the pin does not entirely pass through the end of the striker, there is danger of igniting the fuze if you try to reinsert it.

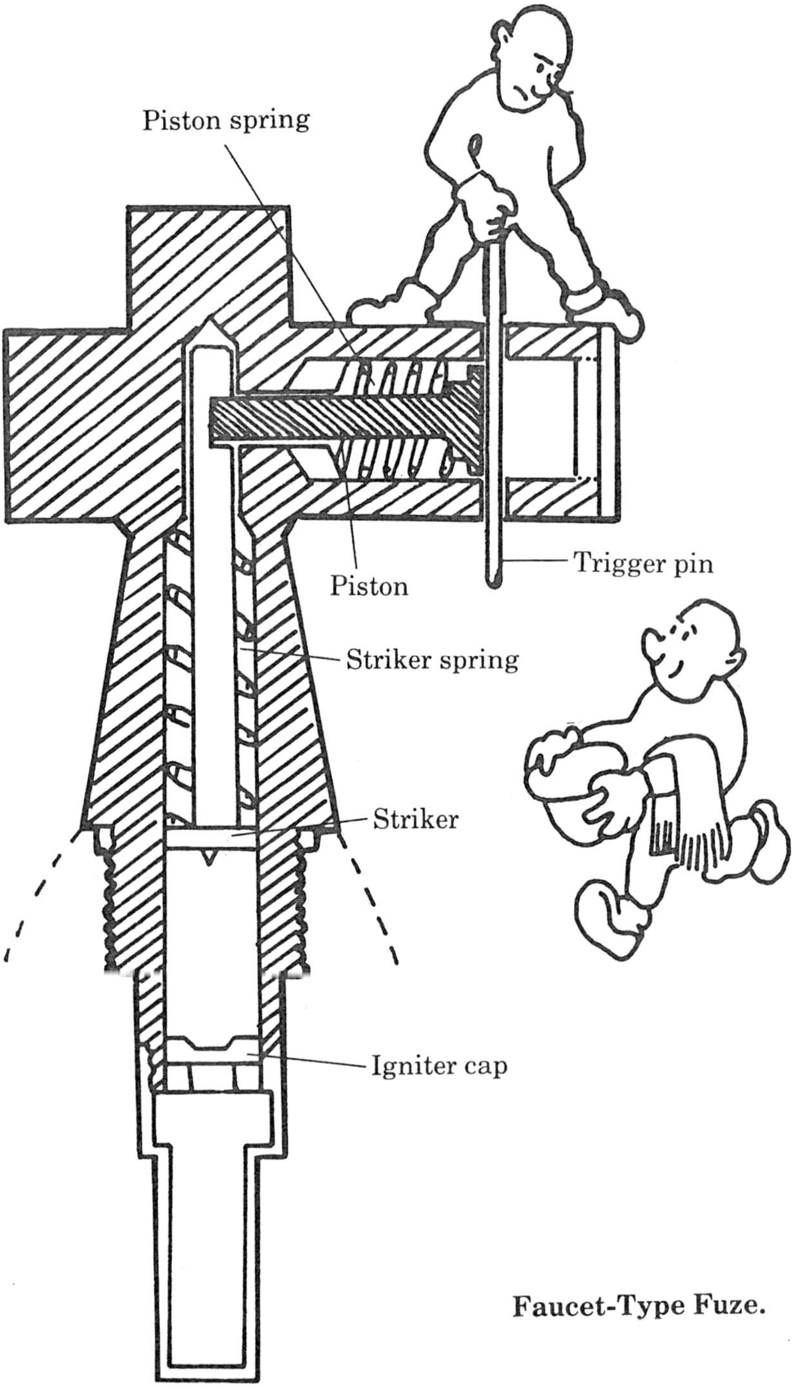

Faucet-Type Fuze.

FAUCET-TYPE FUZE

Description: Resembles a brass faucet. Screw threads enable it to be attached to the ogive of a shell.

Functioning: In the armed position, the striker is pulled back; the spring is held compressed by a piston perpendicular to its axis. The end of the piston fits into a notch in the striker. This piston is held in this position, with the spring compressed, by the trigger pin. A pull on the trigger pin releases the piston, which disengages itself from the notch in the striker. The striker is freed and strikes the igniter cap.

Neutralizing: Cut the pull cord. Take out the fuze.

Note: If the trigger pin does not pass completely through the body of the fuze, do not attempt to reinsert it. Destroy the fuze in place.

Source: Company of the Corps of Engineers 61/3, Tonkin. 1948.

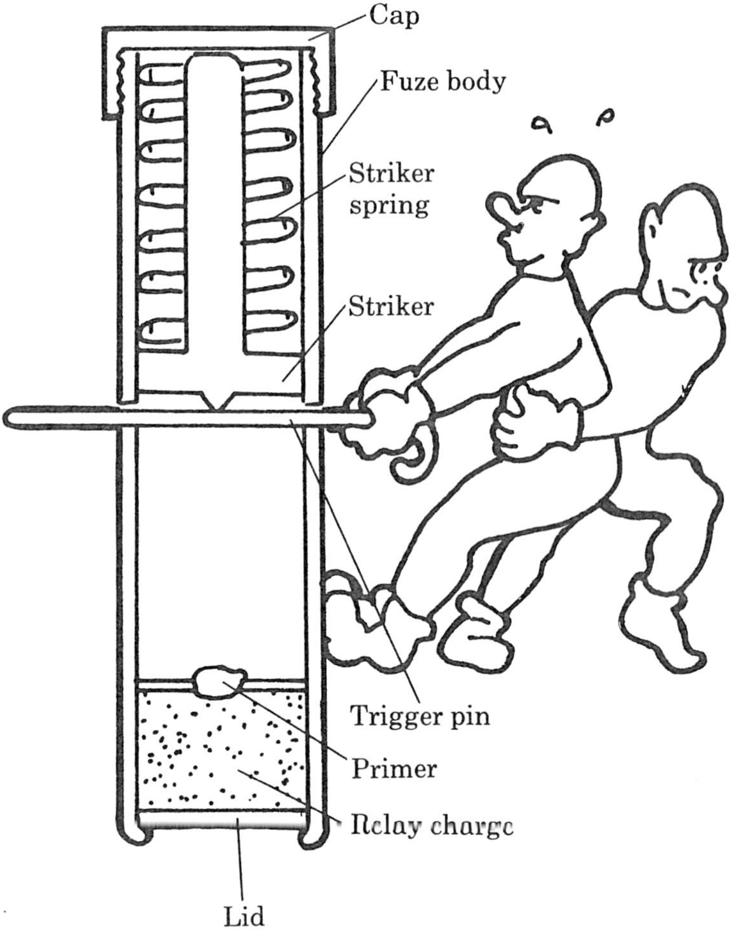

VC Improvized Igniter. Scale: 1:1.

VC IMPROVISED IGNITER

Type: Starter pin.

Appearance: The body of the detonator is a piece of iron water-main pipe with a 30-mm outer diameter.

Functioning: In the armed position, the striker spring is compressed between the head of the striker and the igniter cap, the end of the striker acting like a slide-rod. The striker is held in place by the starter pin. A pull on the pin frees the striker which hits the primer.

To set: The explosive part of the fuze is slid into the chamber of the shell or bomb being used as a mine.

Neutralizing: Not possibly safely. Slowly remove the detonator after cutting the pull cord. Stay at a distance and pull on the pull cord. Beware of flying fragments.

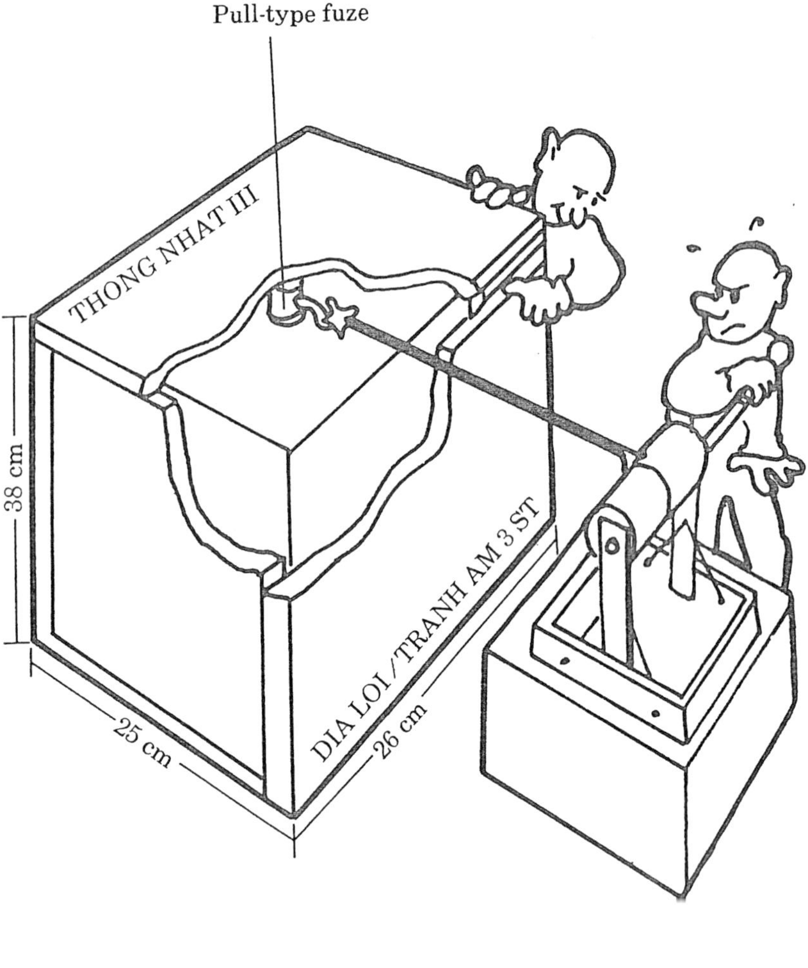
Pull-type fuze
THONG NHAT III
DIA LOI / TRANH AM 3 ST
38 cm
25 cm
26 cm

Boxed Mine.

BOXED MINE

Appearance: A metal mine is placed in a wooden box. with a slit on one side to allow for passage of the pull cord.

Application: Used with a pull cord placed in a booby trap or by remote control with a firing device. May also be used under water.

Dimensions: 38 x 25 x 26 cm.

Inscriptions: On the box: THONG NHAT III (Translation: "Union of 3 Workshop"); DIA LOI/TRANH AM 3 ST (Translation: "Insulated mine").

Functioning: An insulated fuze is set off by a pull on the safety pin.

Neutralizing: Cut the booby-trap wire. If necessary, carry the mine away *inside its case* and destroy. Never attempt to remove the wooden case to "take a peek." Some mines of this type are booby trapped with a grenade with its safety pin removed.

Source: Service des Munitions (Munitions Service), March, 1948.

Pull-Release-Type Fuze. Scale: 1:2.

PULL-RELEASE-TYPE FUZE

Type: Pull-type using a hold-down trigger. Release-type acting by sudden release of a hammer.

Appearance: Parallelepiped. Made of thin sheet metal. Visible end of the hammer is hook shaped.

Functioning: (1) Pull-type: In the armed position, the hammer is retained by the trigger, which is locked inside the catch of the weapon. A pull on the end of the trigger releases the hammer, which springs up, strikes the heel of the firing pin, and breaks its protection spring. (2) Release-type: In the armed position, the mine is suspended from a wire attached to the hook on the end of the hammer. The weight of the mine is generally sufficient to compress the hammer spring. When a vehicle hits the mine, the suspension wire breaks and the hammer strikes the heel of the firing pin.

Note: In the release-type, the trigger is not used.

Neutralizing: Immobilize the hammer. Remove the firing pin. Remove the mine. It is preferable to destroy this very sensitive device by pulling the control wires from a distance.

Source: Cochin-China, 1948. Instruction Center of the Corps of Engineers.

Pivot-Plate Fuze.

PIVOT-PLATE FUZE

Type: Pull-type.

Appearance: Copper hood with a slit on its top perpendicular to the axis of the striker. Inside this slit is an anchor plate. The visible end of the striker passes through the upper part of the hood.

To set: Place the fuze inside a 75-mm shell. Attach the pull cord. Remove the safety pin.

Functioning: In the armed position, the striker spring is compressed. The striker is held back by the edges of the slot in the pivot plate which are inside the retainer groove. A pull on the lever of the pivot plate causes it to rotate around the pivot screw. The striker is released and hits the primer.

Neutralizing: Cut the pull cord. Insert a safety pin, *holding the plate firmly* inside the slit at the top of the hood. Remove the fuze. Be careful, as this is a very sensitive device.

Use: Antivehicular or antipersonnel.

Source: T.F.I.S., 1948.

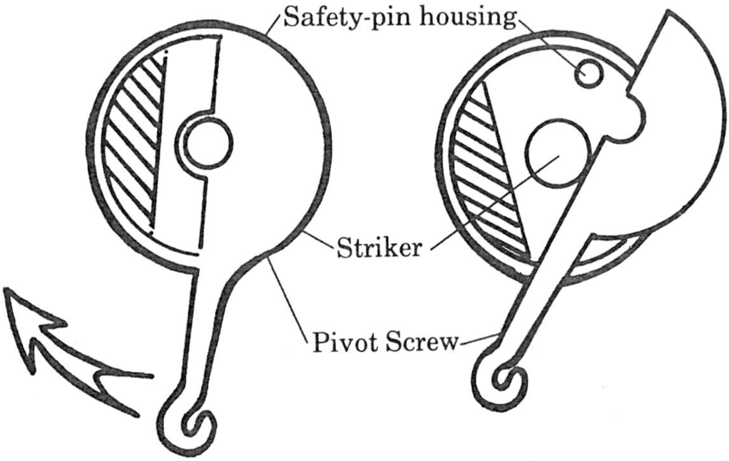

Armed position

Firing position

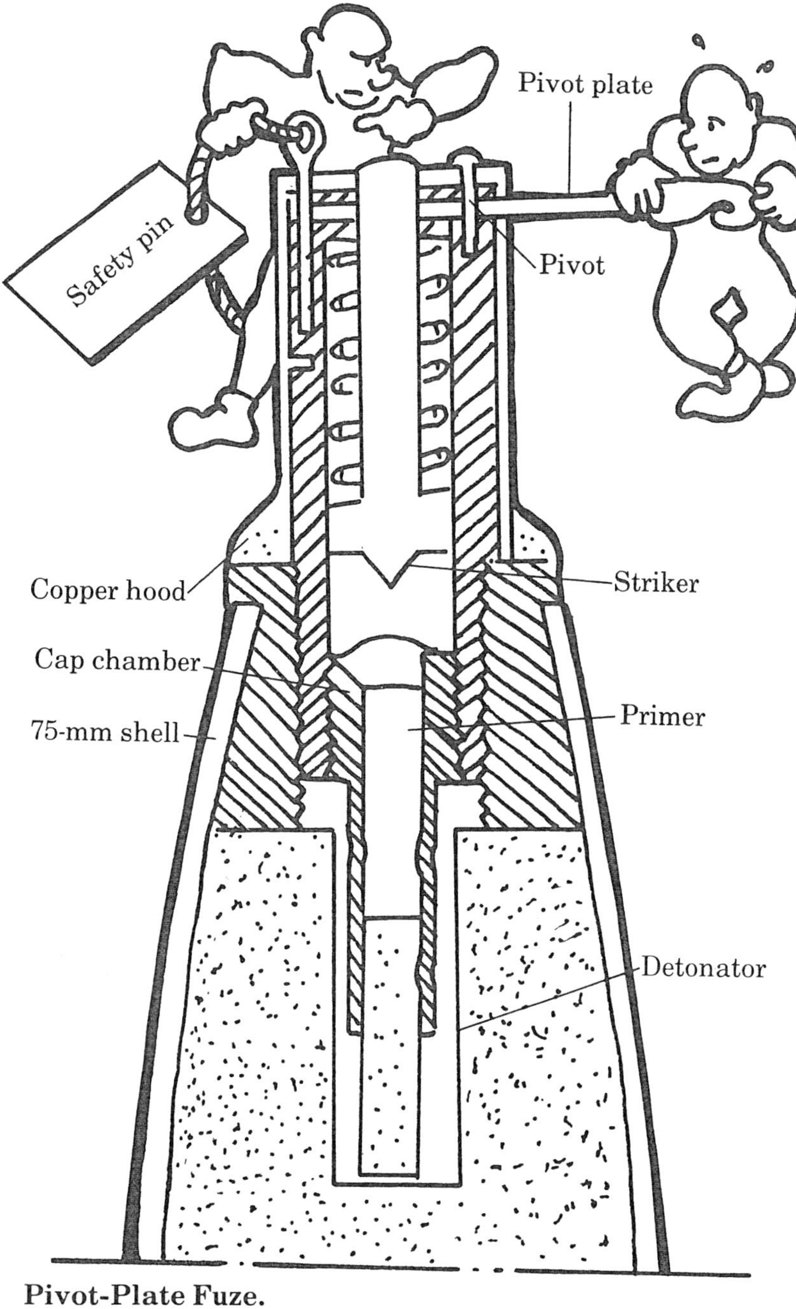

Pivot-Plate Fuze.

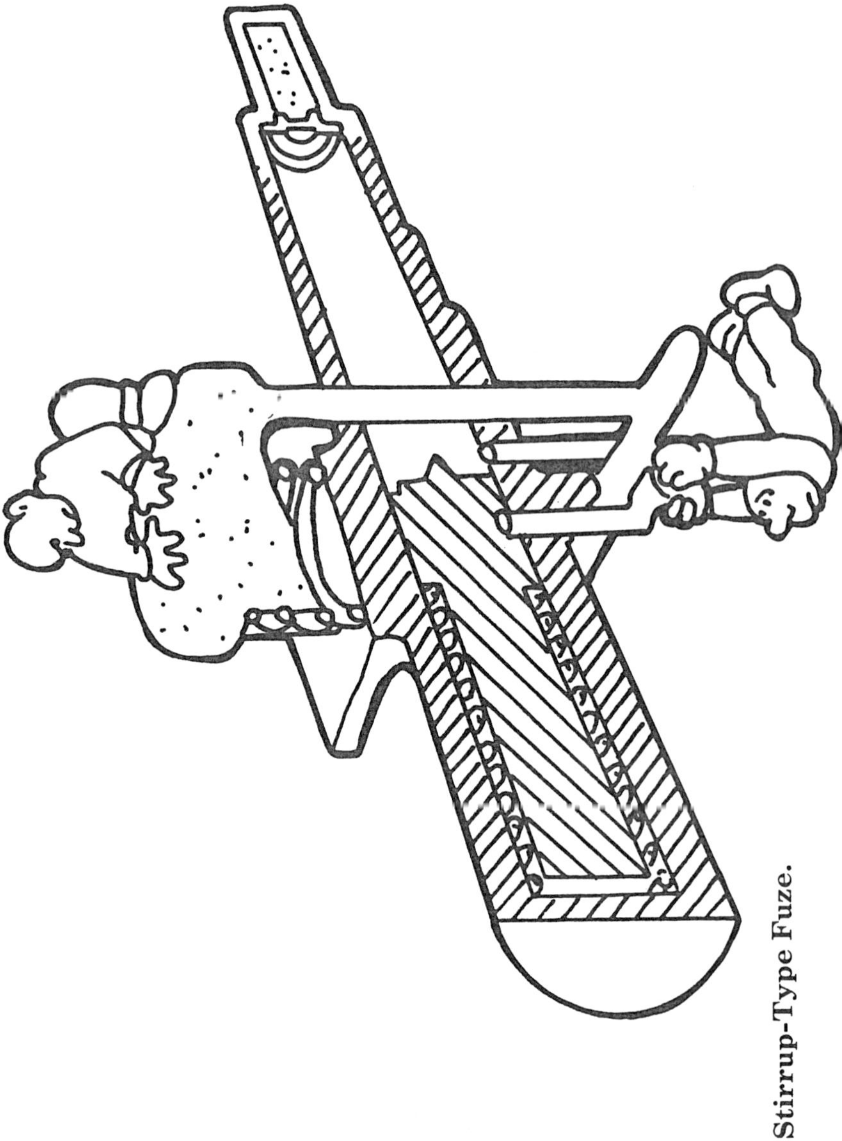

Stirrup-Type Fuze.

STIRRUP-TYPE FUZE

Description: The body of the fuze is cylindrical with a rectangular plate welded perpendicular to its axis at half its length. This plate has two holes for the straps of the stirrup. A trigger spring is compressed between the rectangular plate and the upper flat part of the stirrup.

Functioning: In the armed position, the striker is held back. The spring is compressed by a U-shaped trigger pin, one side of which is slightly longer than the other. The base of the stirrup is in the curved part of the U.

In the firing position: (a) Normal functioning: Pressure on the plate on the stirrup compresses the trigger spring. This pressure is transmitted to the trigger pin which then comes out. The striker is freed and hits the primer. (b) Antidemolition: The U-shaped pin contains a ring to which a pull cord can be attached. Picking up the mine-fuze causes the pin to come loose and the device to explode.

Neutralizing: No safe way exists. It is preferable to destroy the device. In case of absolute necessity, immobilize the pin and stirrup and remove the detonator.

Source: Tonkin, 1947.

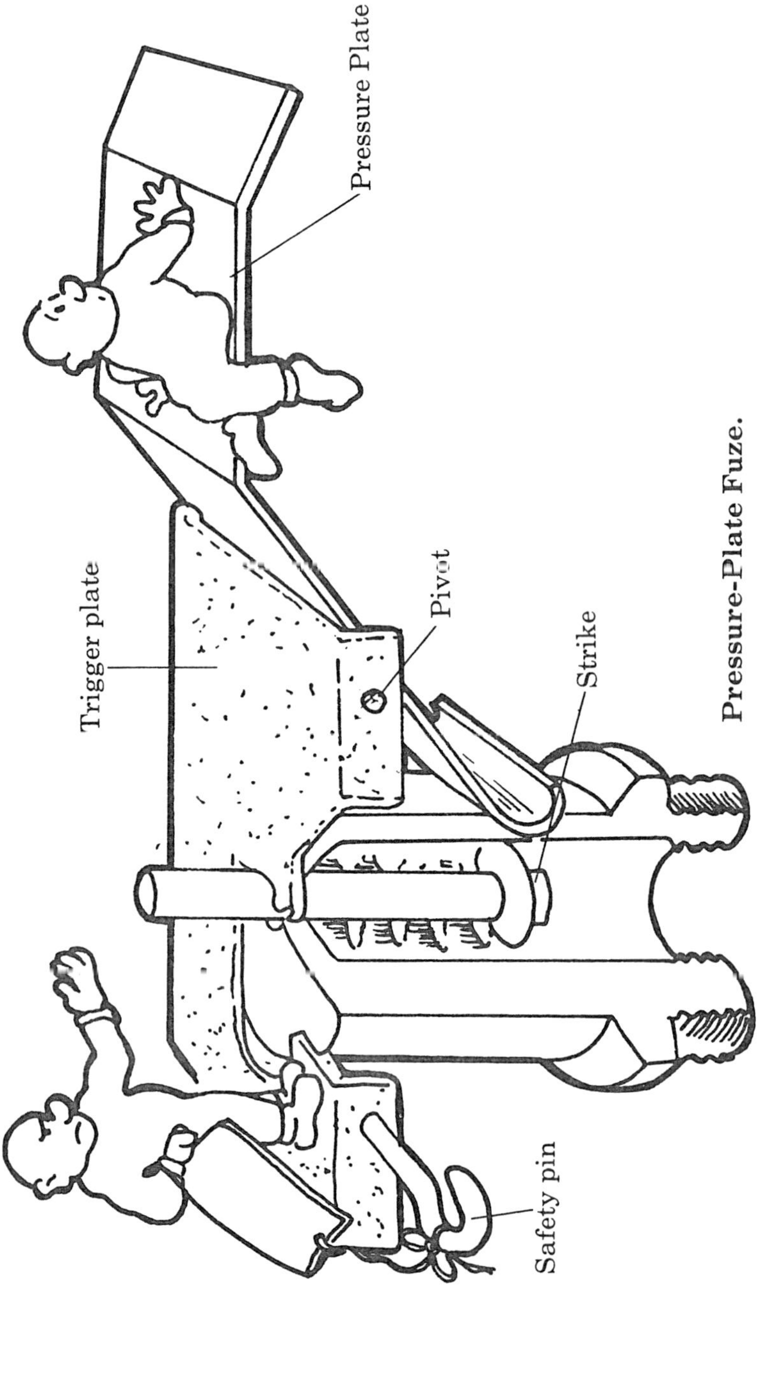

Pressure-Plate Fuze.

PRESSURE-PLATE FUZE

Description: The fuze body is cylindrical in shape (4.5 cm long, 3 cm outer diameter) and made of aluminum alloy. Threading makes it adaptable for use on a Japanese antitank mine.

Functioning: In the armed position, the striker is retained. The spring is compressed by a trigger plate which engages a notch-hole in the striker. Pressure on the outside plate acts as a lever and causes the trigger plate to move and free the striker, which hits the primer.

Neutralizing: Reinsert the safety pin.

Note: If the trigger plate does not appear to be clearly locked inside the notch-hole of the weapon, destroy on the spot.

Source: City of Hanoi, July, 1948.

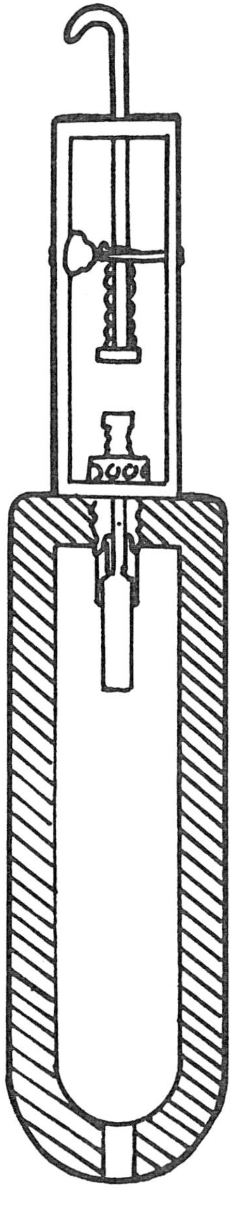

To arm

Mine with Pull-Release-Type Fuze. Scale: 1:2.

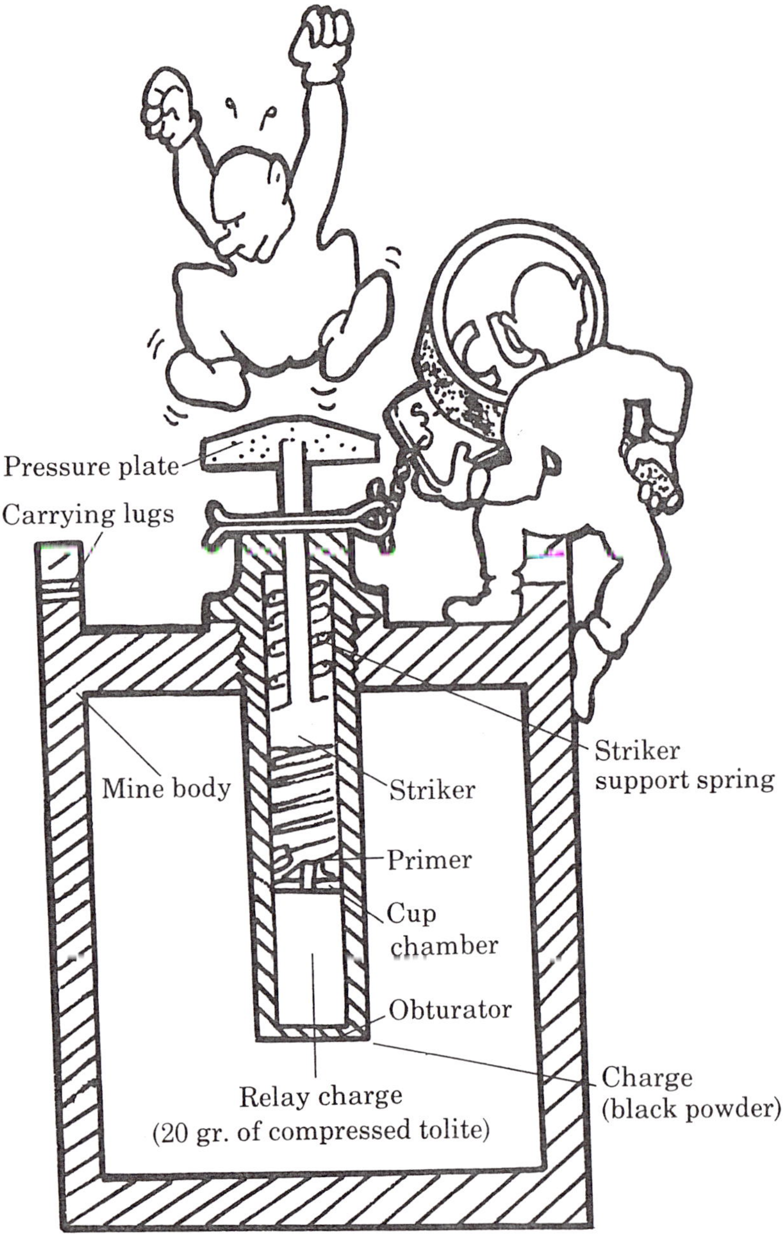

Antivehicular Mine, C.B.X.Q.H. Scale: 1:2.

ANTIVEHICULAR MINE C.B.X.Q.H.

Appearance: Cast-iron cylinder with carrying lugs on top.

Total weight: 15.55 kg.

Explosive charge: 2.5 kg of mixture of black powder and red phosphorus.

To set: Holding the striker, remove the safety pin. Lower the striker slowly into the body of the fuze. Bury the mine in a vertical position with the fuze on top.

Caution: Do not push down on the striker accidentally.

Functioning: In the armed position with the safety pin removed, the striker is held in equilibrium between the two opposing protection springs. Pressure exerted on the small plate covering the end of the striker compresses the lower protection spring, causing the tip of the striker to strike the primer.

Neutralizing: If necessary, keep the mine in the hole and slowly pull on the end of the striker until the opening in the safety pin is visible. Reinsert the safety pin. Pull the mine out, but from a distance and under shelter.

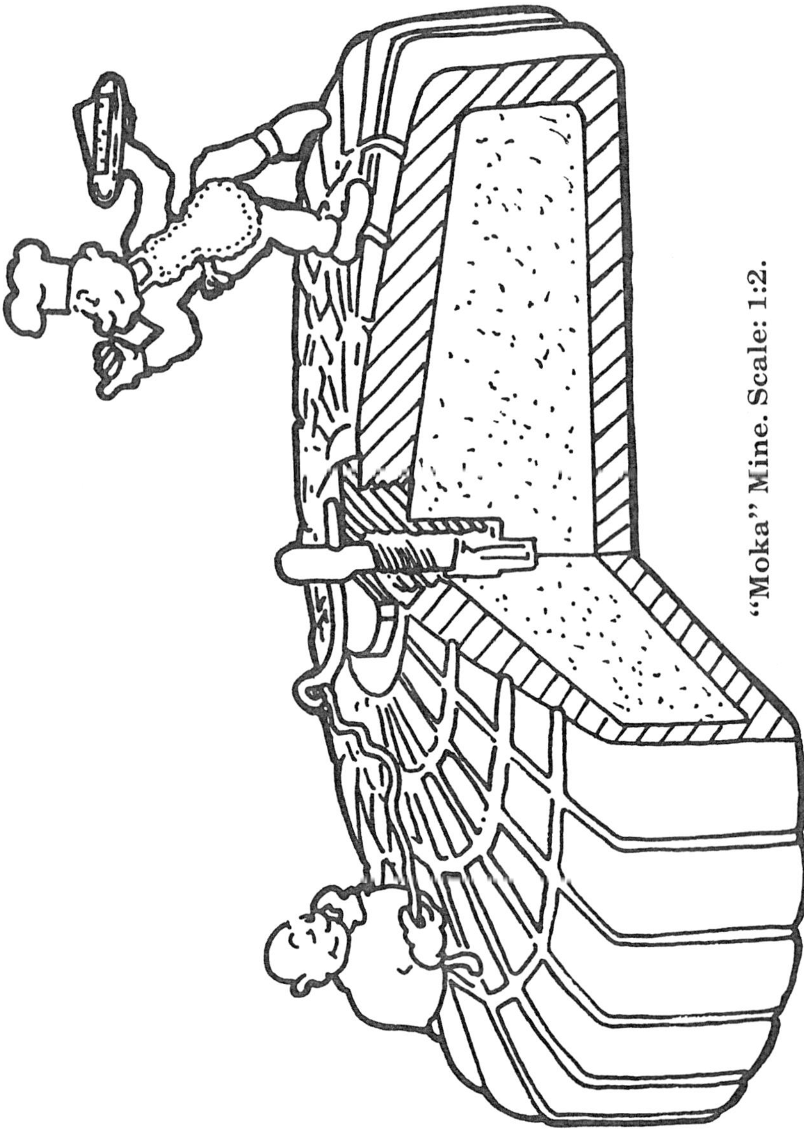

"Moka" Mine. Scale: 1:2.

"MOKA" MINE

Type: Fragmentation; antivehicular and antipersonnel.

Application: Antivehicular: Bury at a shallow depth in the probable path of the vehicles to be destroyed. Antipersonnel: Bury at a shallow depth in the edge of a slope.

Appearance: Circular in shape, similar to German antitank mine. Thick cast-iron body with serrations. Gray-green color. Rudimentary firing device is of brass, inserted into the mine's central chamber.

Functioning: The fuze operates by a pull on a cord wrapped around the retainer pin of the striker. This cord is placed along a small groove in the ground.

Neutralizing: Cut the pull cord. Pull on it from a distance (beware of a possible grenade buried as a booby trap underneath the mine). If the pull pin does not clearly pass through the end of the striker, destroy the mine in place. If the mine explodes above ground level, the fragments can be dangerous within a radius of 200 meters.

Source: Annam, 1948. Instruction Center of the Corps of Engineers.

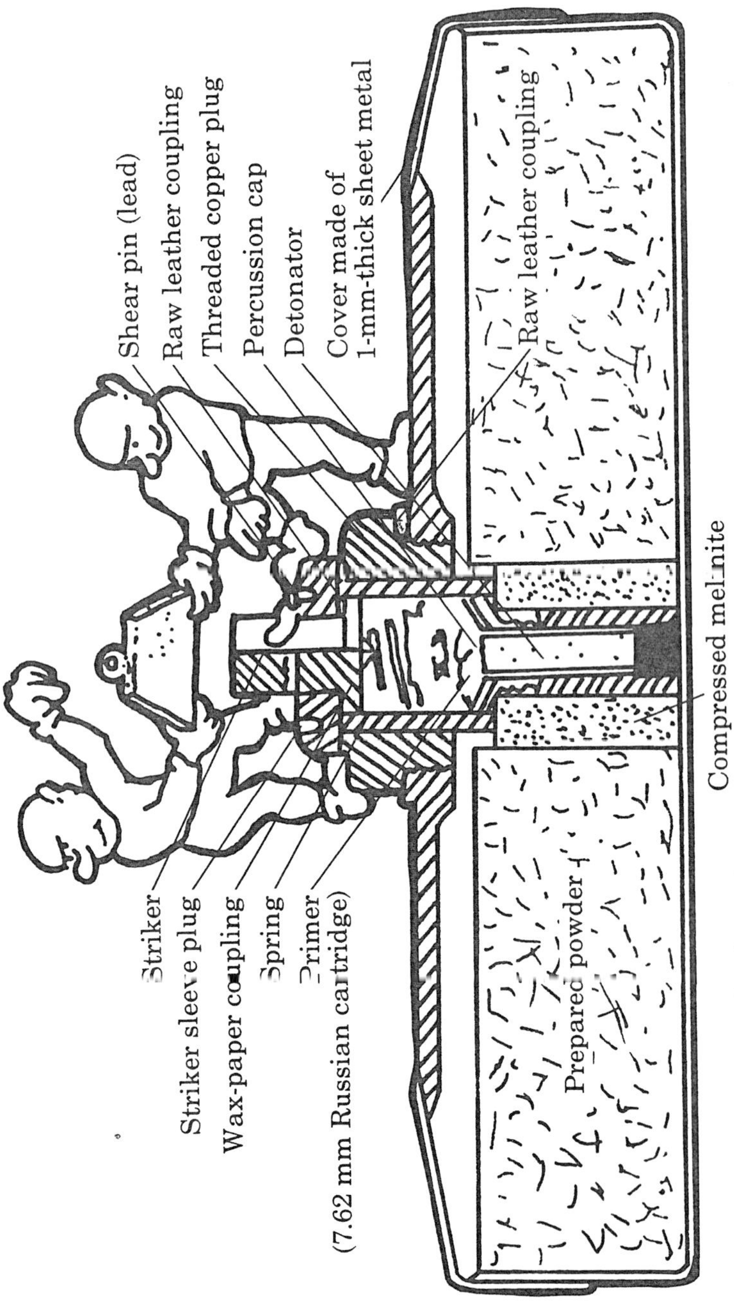

Mine with Lead Pin. Scale: 1:1.

MINE WITH LEAD PIN

Appearance: "Saucer" mine. Central fuze. A shear pin is visible passing through the upper part of the fuze.

Functioning: In the armed position, the striker is held by a lead shear pin and a protection spring compressed between the head of the striker and the cap chamber. Downward vertical pressure on the tip of the striker shears the lead pin. The conical protection spring is compressed and the percussion cap is broken, setting off the explosion.

Neutralizing: Take out the fuze slowly. In case of misfire, destroy the mine in place.

Origin: 1948 information.

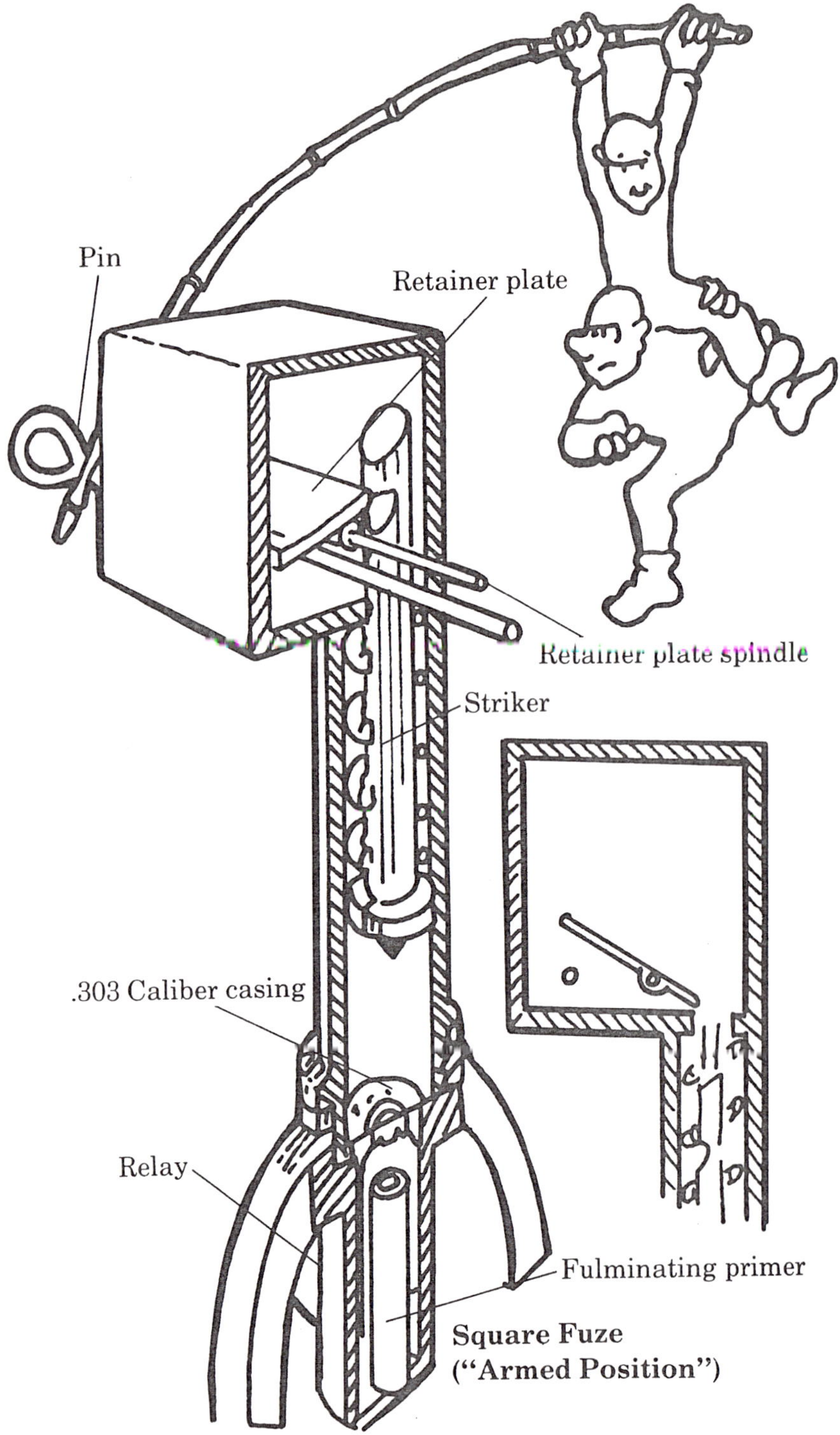

Square Fuze
("Armed Position")

SQUARE FUZE

Type: Small plate acting as a retainer trigger.

Appearance: Square box with a safety pin passing through it. The box is welded onto a tube that makes up the body of the fuze.

To set: Place the fuze into the chamber of the shell or bomb used as the mine. Attach the pull cord to the safety pin.

Functioning: In the armed position, the safety pin immobilizes the retainer plate. The tip of the striker has a notch hole which the retainer plate engages. The striker spring is compressed. A tug on the cord pulls out the safety pin. The plate pivots on its spindle and releases the striker, which hits the primer.

Neutralizing: Cut the pull cord. If the safety pin clearly passes through the box, lift out the fuze. Otherwise, destroy the device on the spot.

Source: VC document, 1948.

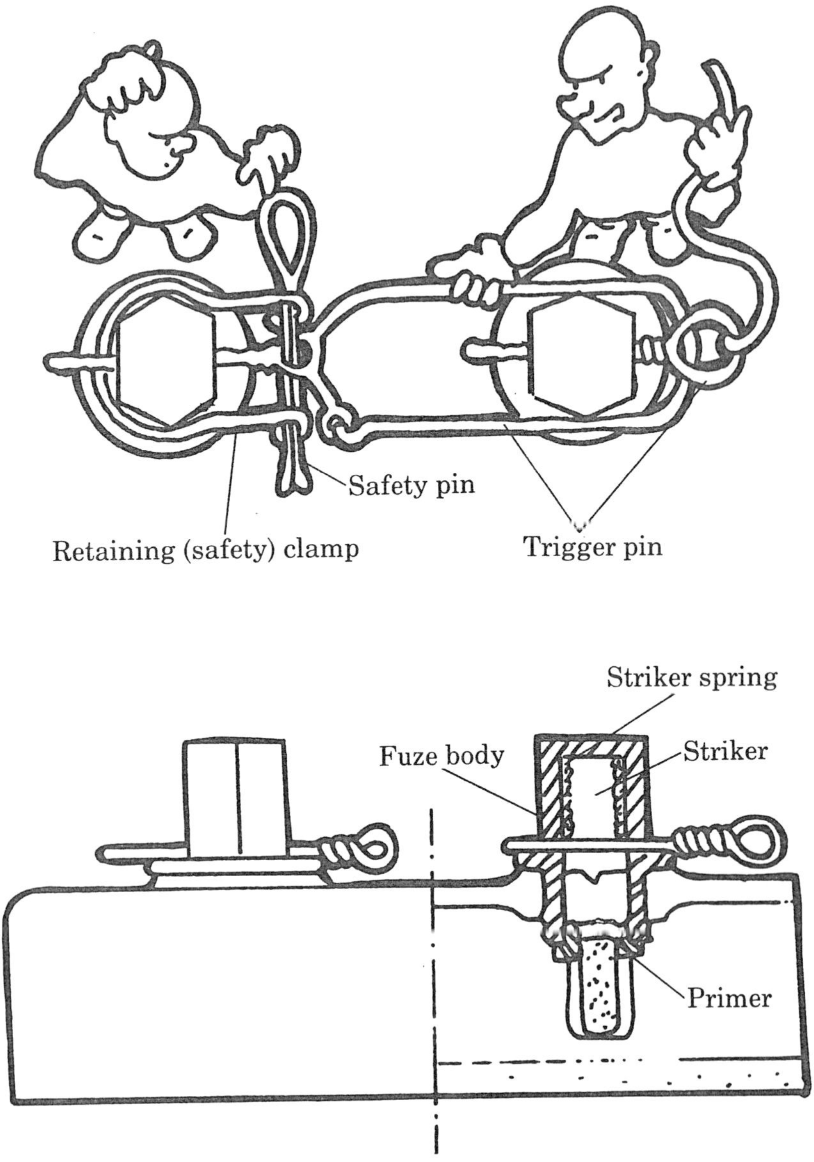

VC Flat Mine With Two Pull-Type Fuzes. Scale: 1:2.

FLAT MINE WITH TWO FUZES

Type: Antivehicular.

Current employment: Road holes.

Appearance: Circular in shape. Made of 4-mm thick welded and stamped iron. Two threaded round orifices on the top are used for filling. The fuzes are placed into these orifices.

Explosive charge: Approximately 2.5 kg of black powder.

Firing device: Two identical fuzes with a common trigger-pin pull system.

To set: Place the mine in the hole. Attach the pull wire to the pull ring. Remove the safety pin. Camouflage.

Functioning: A tug on the wire simultaneously pulls out the two trigger pins. The strikers are released and detonate the primers.

Temporary neutralizing: Cut the pull cord. If the safety clamp is still on, insert the safety pin. If not, bind with a thin piece of metal wire so as to immobilize the two trigger pins.

Permanent neutralizing: Gently clip in two the part connecting the two trigger pins. Remove the fuzes.

Source: Corps of Engineers T.F.I.N. A dispatch boat called "La Gazelle" (tied to a junk in the vicinity of Tonkin).

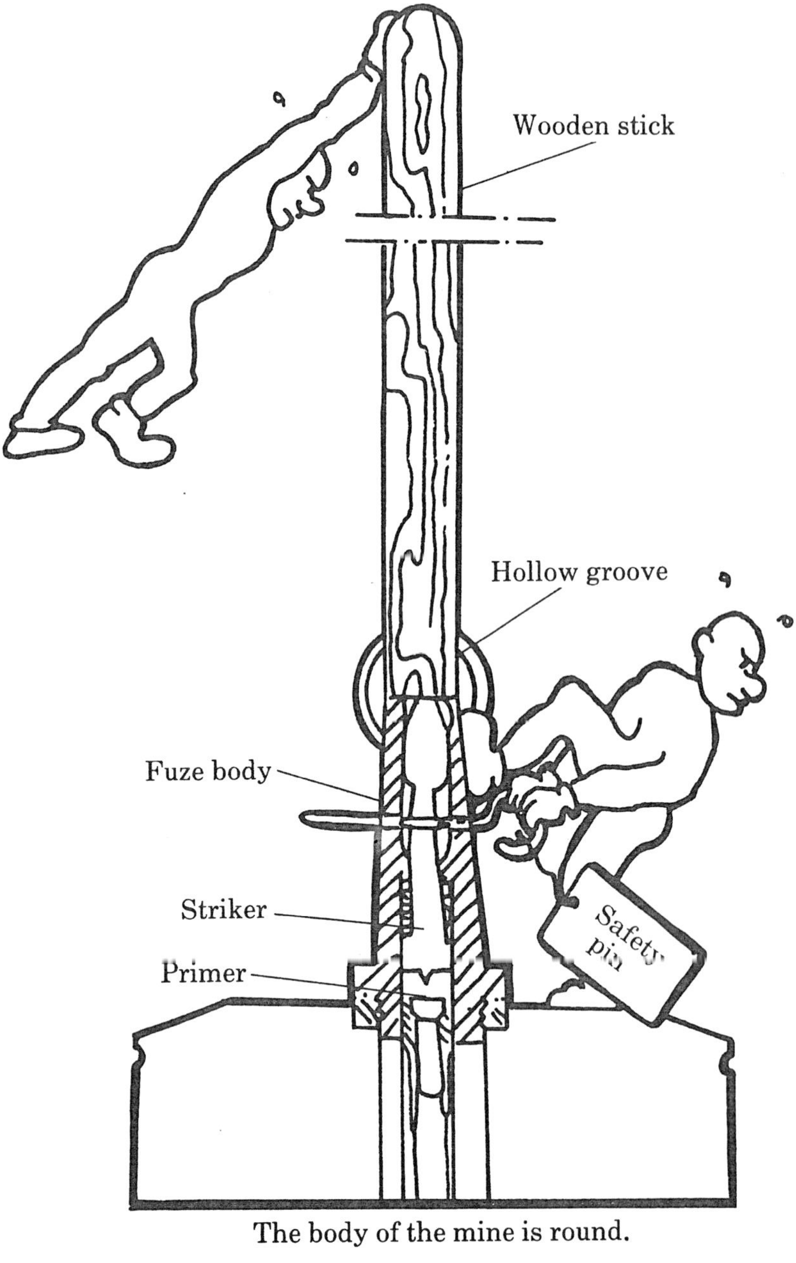

The body of the mine is round.

Bamboo Mine. Scale: 1:2.

BAMBOO MINE

Type: This mine contains a stick that is ejected and pierces an object.

Appearance: The body of the mine is round and flat. The stick is made of a wooden stem 2 cm in diameter and 50 cm long. It does not appear that the small diameter of the mine (17 cm) and its light weight would give sufficient stability. Some sort of weight must be used to steady the base.

To set: Camouflage the stick in underbrush. Steady the base. Remove the safety pin.

Functioning: In the armed position, the striker spring is compressed by the striker. The end of the striker is placed inside the base of the stick. The shoulder of the hollow groove sticks into the upper part of the fuze body and opposes the action of the striker spring. Movement that tends to bend the stick in any direction causes it to break at the level of the hollow groove. The striker is freed and hits the primer.

Neutralizing: Reinsert the safety pin. Remove the fuze.

Caution: The stick may have deteriorated as a result of termites or atmospheric conditions. It is always *very fragile.*

Current usage: The mine is placed in underbrush or a bamboo grove. The stick is often covered with a piece of bamboo.

Source: T.F.I.S., 1948.

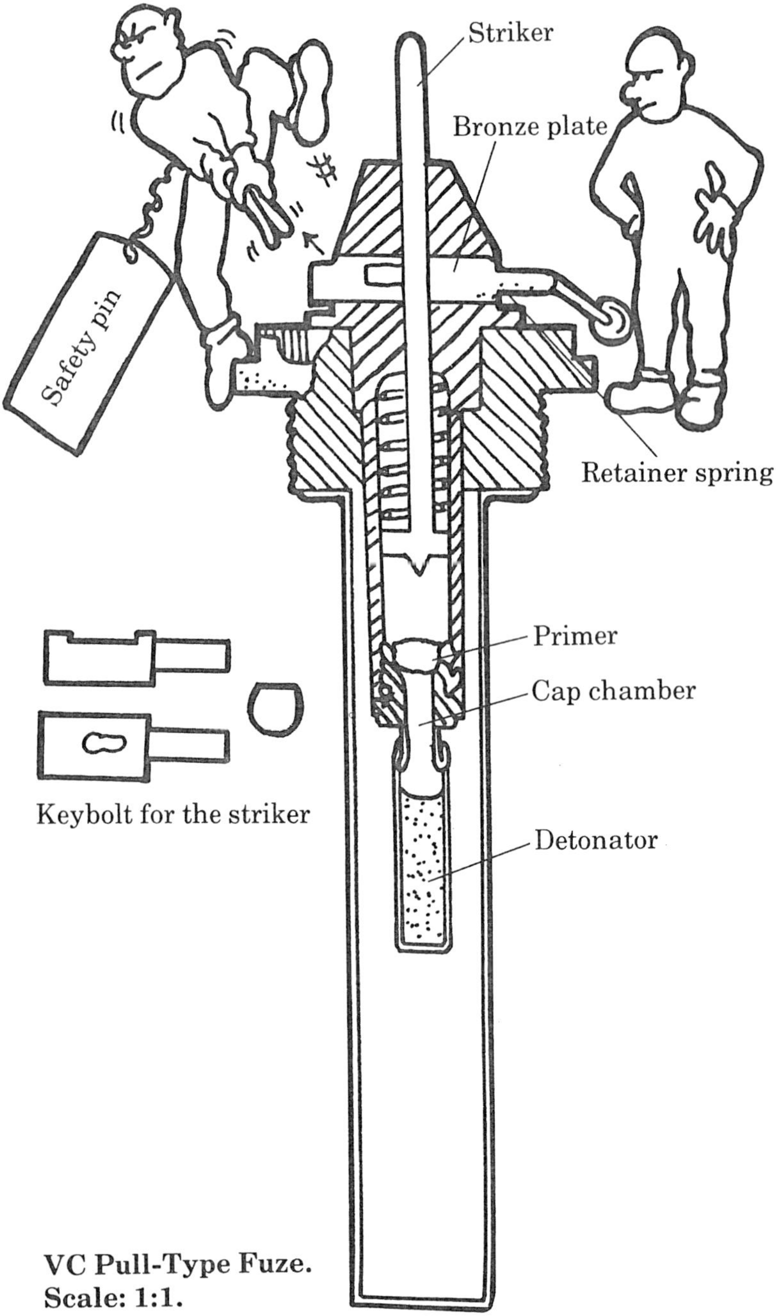

VC Pull-Type Fuze.
Scale: 1:1.

VC PULL-TYPE FUZE

Type: With keybolt

Functioning: In the armed position, the striker spring is compressed and is held by the keybolt, the notched part of which fits into the narrow end of the keyhole. This is held in position by a retainer spring. A pull on the bow of the key compresses the retainer spring. The key moves and the circular part of the keyhole reaches and releases the striker, which strikes the primer.

To set: Insert the cap chamber. Insert the fuze into the mine. Attach the pull cord (unwound) to the pull ring. Remove the safety pin from a distance.

Neutralizing: Reinsert the safety pin. Cut the pull cord. Take out the fuze.

Source: Instruction Center of the Corps of Engineers of the Far East, Cochin-China, August, 1948.

Underwater Mine with Two Pull-Type Fuzes.

MINE-BOMB WITH TWO PULL-TYPE FUZES

Type: Smooth-bore bomb.

Appearance: Cast iron. Cylinder shaped with two frustums of cone on each side, the small base parts of which each have a hole in which to insert a fuze.

Dimensions: Outer diameter = 130 mm. Length = 372 mm.

Explosive charge: Approximately 2 kg of explosive similar to black powder.

Firing device: Pull-type fuzes described on the following pages.

Neutralizing: Remove the fuzes after disassembly.

Note: Some of these mines have been used with only one fuze or with an electric igniter.

Source: Instruction Center of the Corps of Engineers of the Far East, August, 1948.

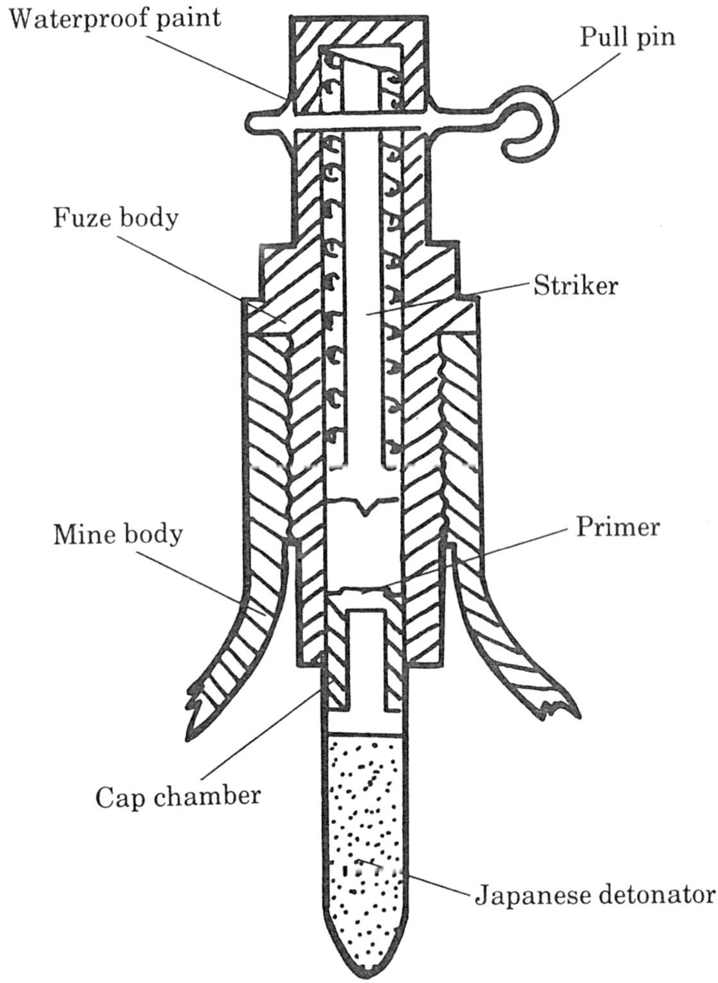

Insulated Fuze for Underwater Mine with Double Igniter. Scale: 1:1.

CAUTION

Do not play with fire.

A mine knows neither friend nor foe.

It waits for the first mistake; each move is final.

You can never be careful enough.

Remember, you can make a mistake only once, and from this mistake you probably won't recover.

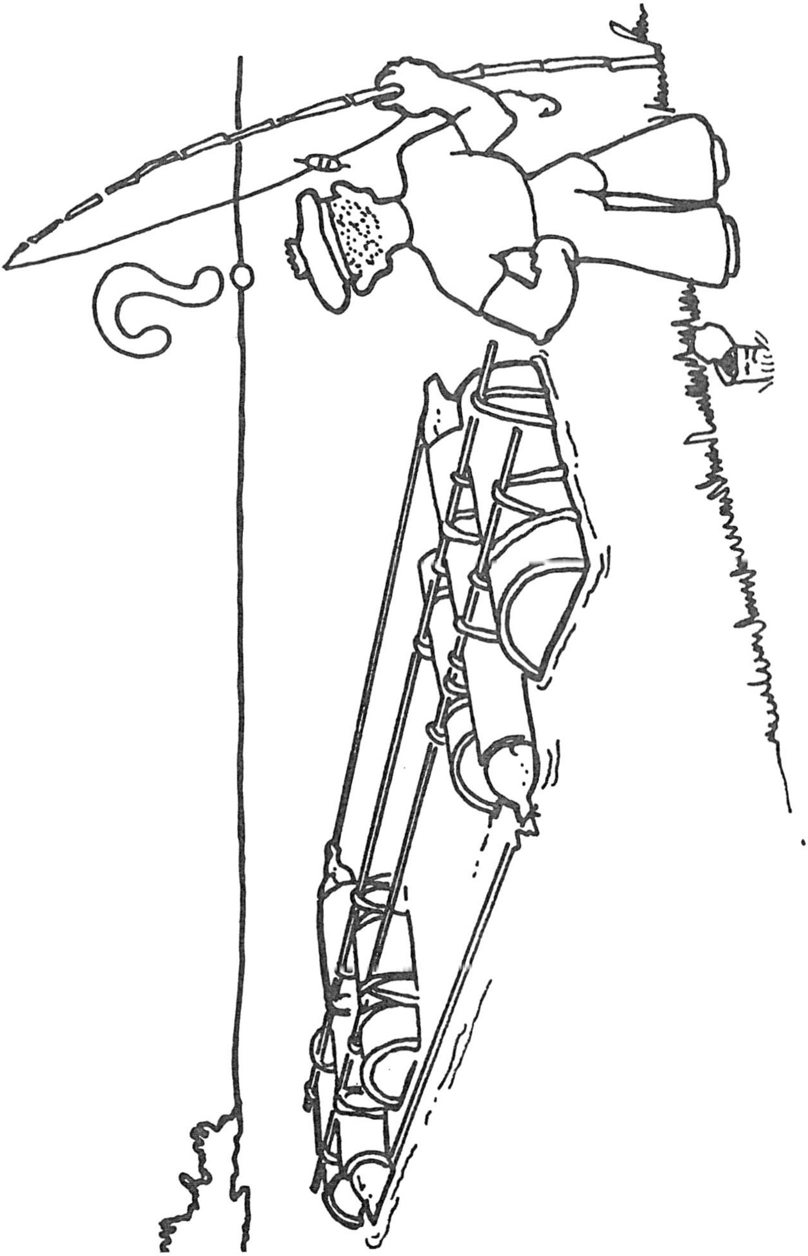

Drifting Mine with Four 280-mm Shells.

CAUTION

Act fast, *but:*

— Do not confuse speed with overhastiness.

— Remember that most mines contain enough explosive to blow off your hand and possibly even kill you.

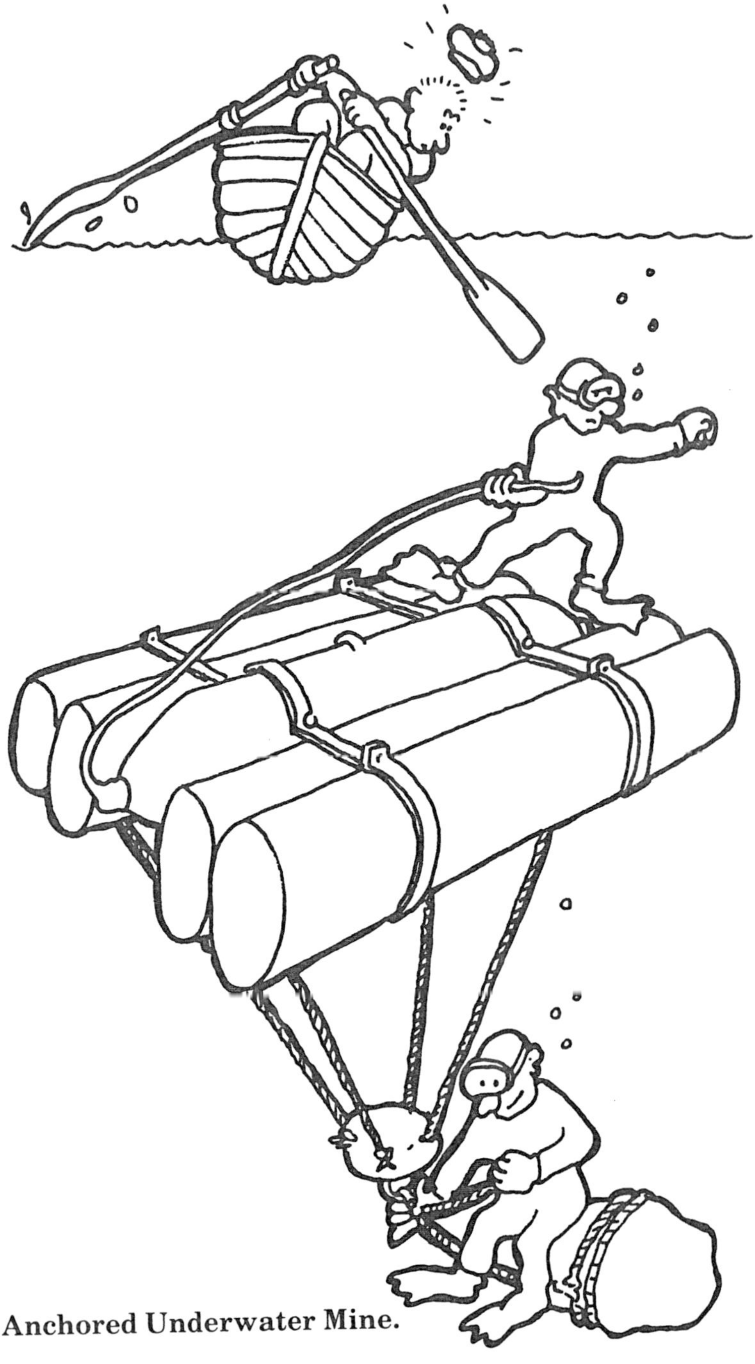

Anchored Underwater Mine.

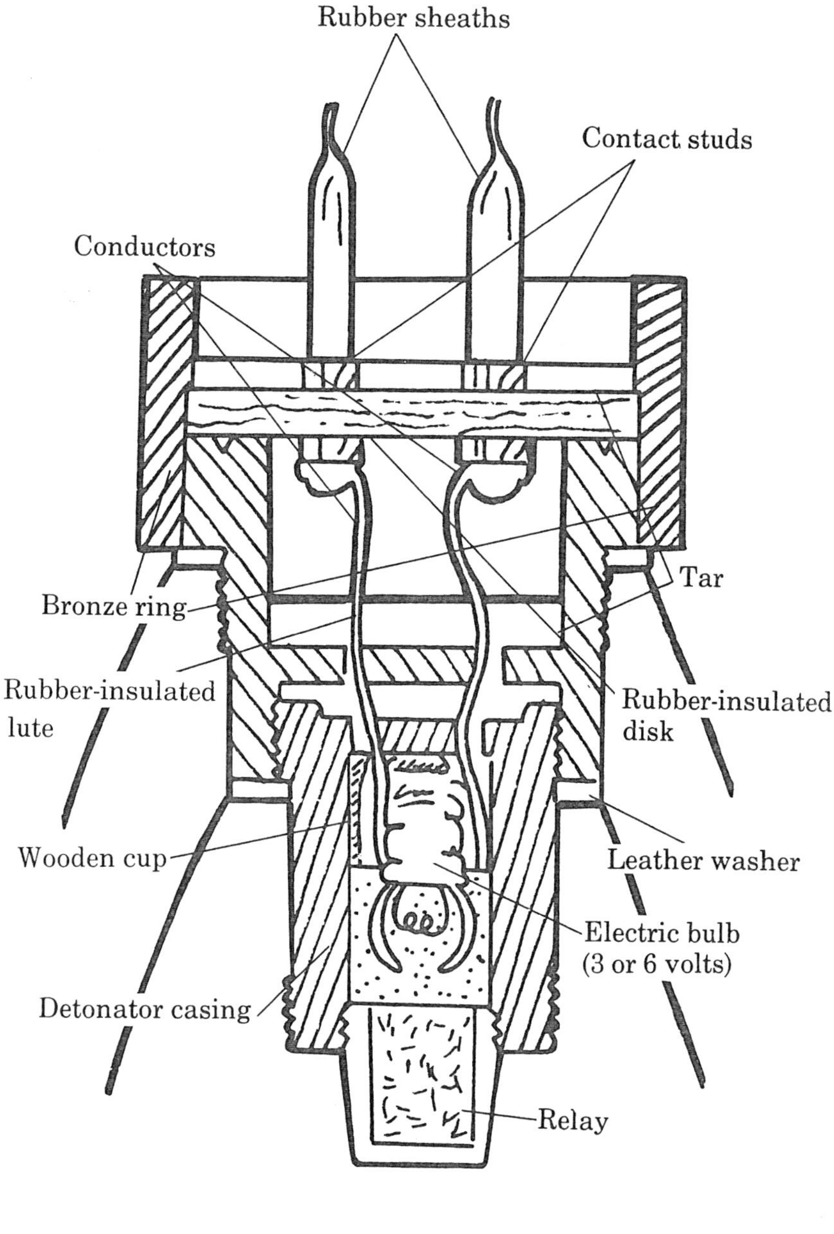

Insulated Electric Fuze for Anchored Underwater Mine.

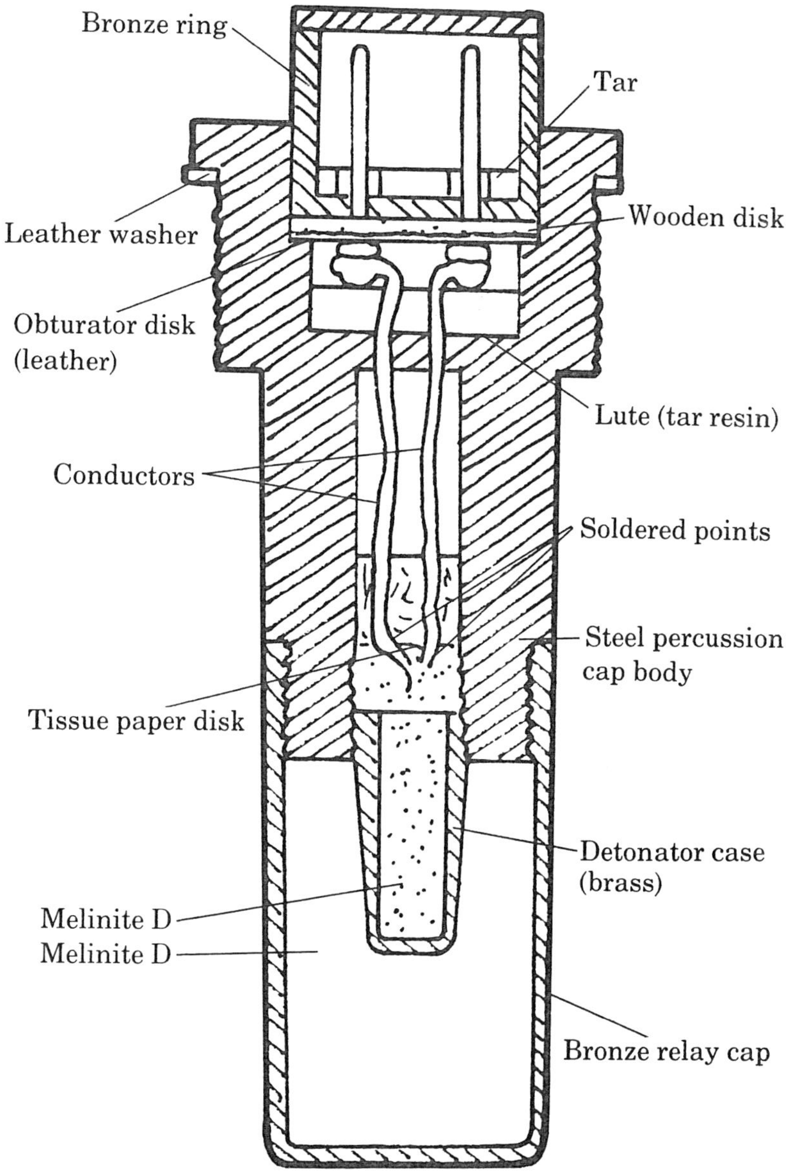

Insulated Electric Fuze for Underwater Mine.

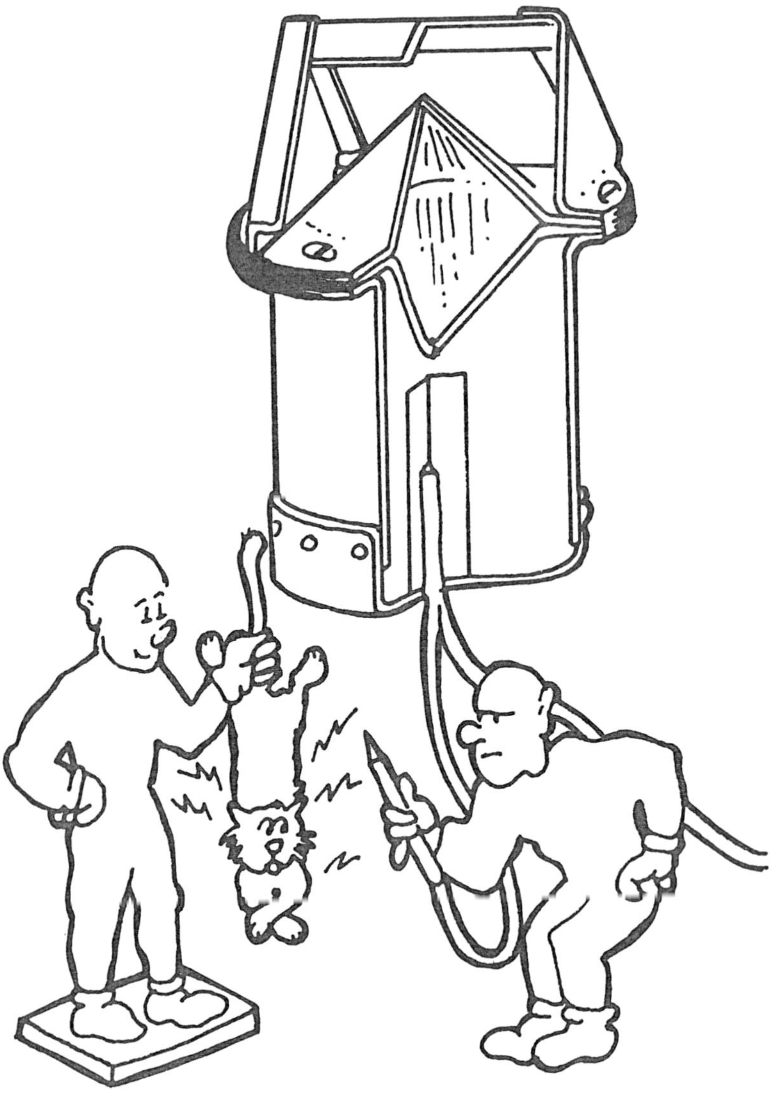

Mine with Hollow Charge (Electric Fuze). Scale: 1:2.

MINE WITH HOLLOW CHARGE I

Type: Hollow charge with a conical recess.

Appearance: Sheet metal, black tar colored. Total weight: 2.8 kg.

Antivehicular model: Weight of explosive: 1.2 kg. Total weight: 4.5 kg.

Antitank model No. I: Weight of explosive: 2.0 kg. Total weight: 14.0 kg.

Antitank model No. II: Weight of explosive: 7.0 kg.

Functioning: When a vehicle passes over it, the mine, which is electrically ignited, explodes. The mine is remotely controlled by an operator using an exploder or a battery.

Note: Antitank model No. II should bring any type of armored vehicle to a stop.

Neutralizing: Cut the pull cords one at a time. Remove the mine from the ground. It can be carried as is with caution. Disassembly operations must be carried out by a specialist.

Source: Cochin-China, 1948. Instruction Center of the Corps of Engineers.

Beehive Mine. Scale: 1:3.

BEEHIVE TYPE HOLLOW CHARGE MINE

Type: Hollow charge with a conical recess.

Appearance: Made of sheet metal with a wooden cap. Total weight: 11.5 kg. Explosive: 6 kg of melted melinite.

Functioning: Electrically ignited by an operator.

To set: As a mine: Bury with base part level with the ground and igniter below. As a demolition explosive: Place base of the mine against the object to be destroyed.

Neutralizing: Cut the lead wires one at a time. Remove the wooden cap.

Note: Although no firing stand that would determine the optimum operating distance has been found, it appears that maximum efficiency can be obtained by placing the mine about 30 cm from the surface of the object to be destroyed.

Source: Cochin-China, 1948. Instruction Center of the Corps of Engineers.

Mine with Hollow Charge (Half-Section). Scale: 1:2.

MINE WITH HOLLOW CHARGE II

Type: Hollow charge with conical recess.

Appearance: Made of sheet metal. Black-tar color. Total weight is 1.4 kg, including about 400 gr of melted melinite.

Functioning: Upon passage of a vehicle, the mine is electrically ignited by an operator using an exploder or a battery. This mine can immobilize a vehicle if it explodes beneath a wheel. If it explodes beneath the chassis, the vehicle can be slightly damaged.

Neutralizing: Cut at least one electric conductor wire. Remove the mine from the ground. Take off the cone head, take out the lining, and remove the electric firing system. It is preferable to have a specialist do this job.

Note: Some mines of this type have a metal bracket or handle attached to the cone. This allows it to be carried and makes it easier to attach the mine to the object to be destroyed.

Hollow Charge with Electric Igniter. Scale: 1:2.

HOLLOW CHARGE MINE WITH BASE STAND

Type: Hollow charge with conical recess.

Appearance: Made of sheet metal. Black-tar color. Total weight: 5.2 kg. Weight of explosive: 2.5 kg.

Functioning: When an armored vehicle passes over it, the mine is exploded electrically by an operator. This explosion can shear the caterpillar track on a tank or perforate the interior armor.

Neutralizing: Cut the conductor wires one at a time. Remove the mine from the ground. It can be carried as is with caution. Disassembly should be carried out by a specialist.

Source: Cochin-China, 1948. Instruction Center of the Corps of Engineers.

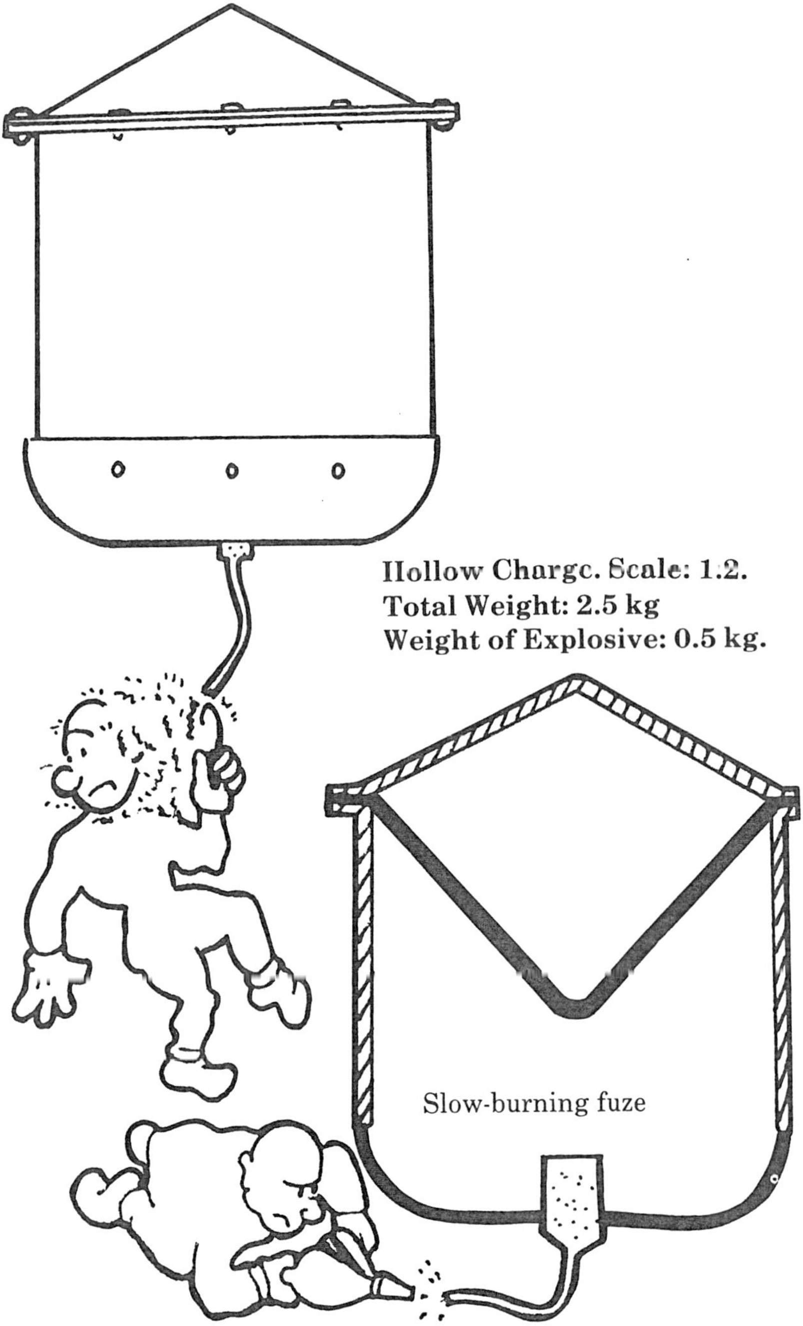

Hollow Charge. Scale: 1:2.
Total Weight: 2.5 kg
Weight of Explosive: 0.5 kg.

VC HOLLOW CHARGE WITH SLOW-BURNING FUZE

Dimensions: This type of mine reportedly exists in several sizes, similar in appearance, varying in explosive charge between 2 and 7 kg.

Functioning: The mine is detonated by a slow-burning fuze of variable length. Models with electric fuzes also reportedly exist. The explosive action occurs inside a hollow charge, and proceeds along the axis of the inner recess.

Neutralizing: If a saboteur has been seen in the vicinity and there is doubt whether the slow-burning fuze has been ignited, stay out of the danger area and wait one hour, if possible, before approaching the mine. If a misfire of the fuze is detected, destroy the mine on the spot. If there has been no ignition, remove the fuze.

Source: 1948 information.

Tripod Hollow Charge. Half-Section. Scale: 1:4.

TRIPOD HOLLOW CHARGE

Appearance: Resembles the beehive mine. It has three legs which regulate the optimum effective distance. There is no feed line.

Dimensions: Outer diameter: 18 mm. Height: 25 mm.

Weight of explosive: 2.1 kg.

Fuze: Pull-type (see following page).

Neutralizing: Remove the fuze gently.

Source: Munitions Service T.F.S.A.P., April, 1948.

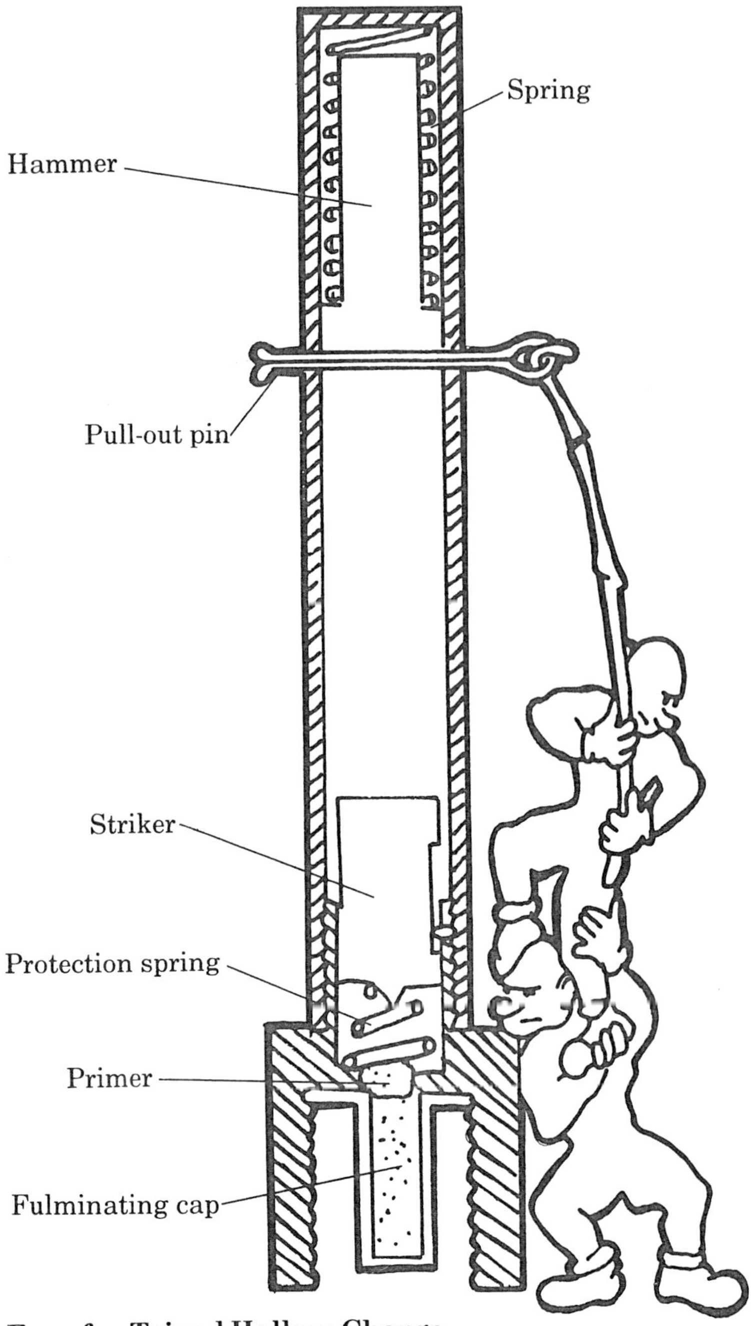

Fuze for Tripod Hollow Charge.

FUZE FOR TRIPOD HOLLOW CHARGE

Appearance: Tube with a pull pin passing through it.

Dimensions: Outer diameter = 20 mm.
Length = 180 mm.

To set: Place the fuze in the upper part of the tripod mine. Attach a pull cord to the safety pin.

Functioning: In the armed position, the trigger spring is compressed by the hammer, which is blocked by the safety pin. The striker is held away from the primer by the protection spring. A pull on the cord releases the safety pin. The hammer is freed and is driven by the operating spring to hit the striker. The striker crushes the protection spring and fires the primer.

Neutralizing: Cut the pull cord and remove the fuze.

Source: Munitions Service T.F.S.A.P., April, 1948.

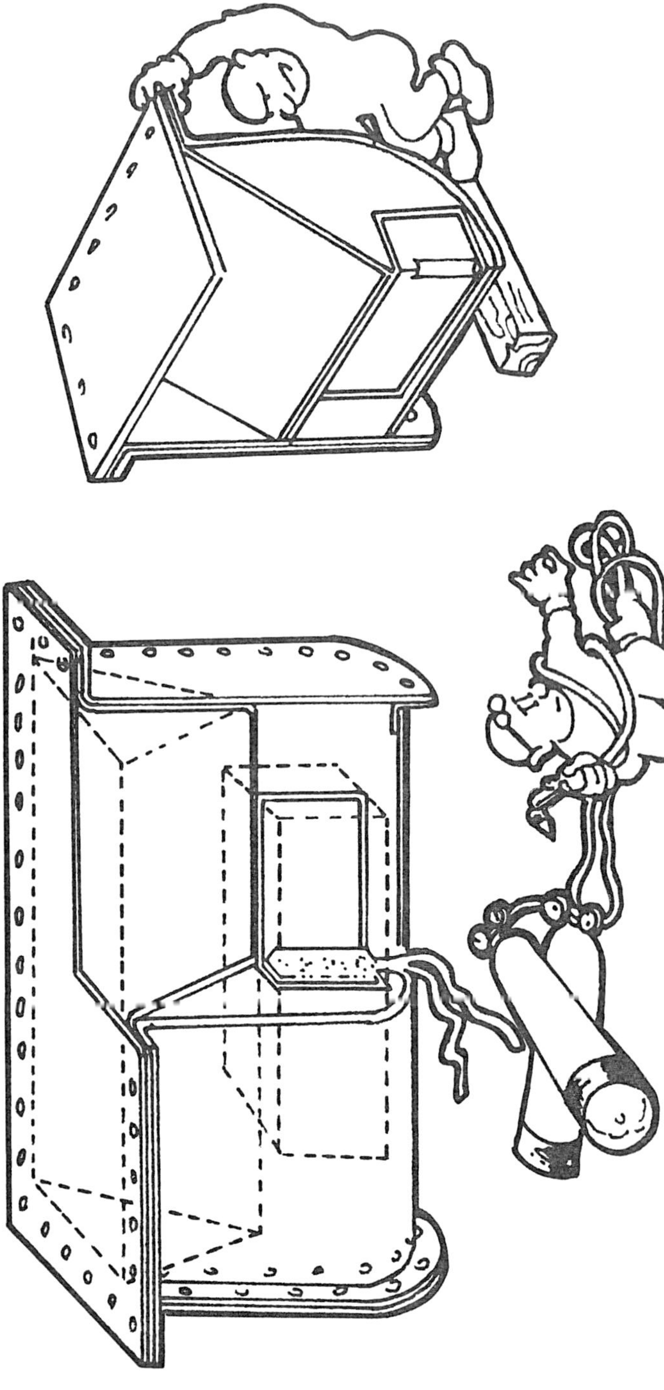

Elongated Hollow Charge. Scale: 1:2.

ELONGATED HOLLOW CHARGE

Type: Hollow charge with prismatic recess.

Appearance: Sheet metal, red-brick color. Total weight: 3 kg. Weight of explosive: 600 gr (melinite).

Functioning: Electrically detonated by an operator.

To set: As a mine: Mine is buried with the rectangular plate even with the surface of the ground. As a demolition explosive: The rectangular plate is placed against the object to be destroyed.

Neutralizing: Cut the conductor wires one by one. The mine can then be removed with caution.

Source: Cochin-China, 1948. Instruction Center of the Corps of Engineers.

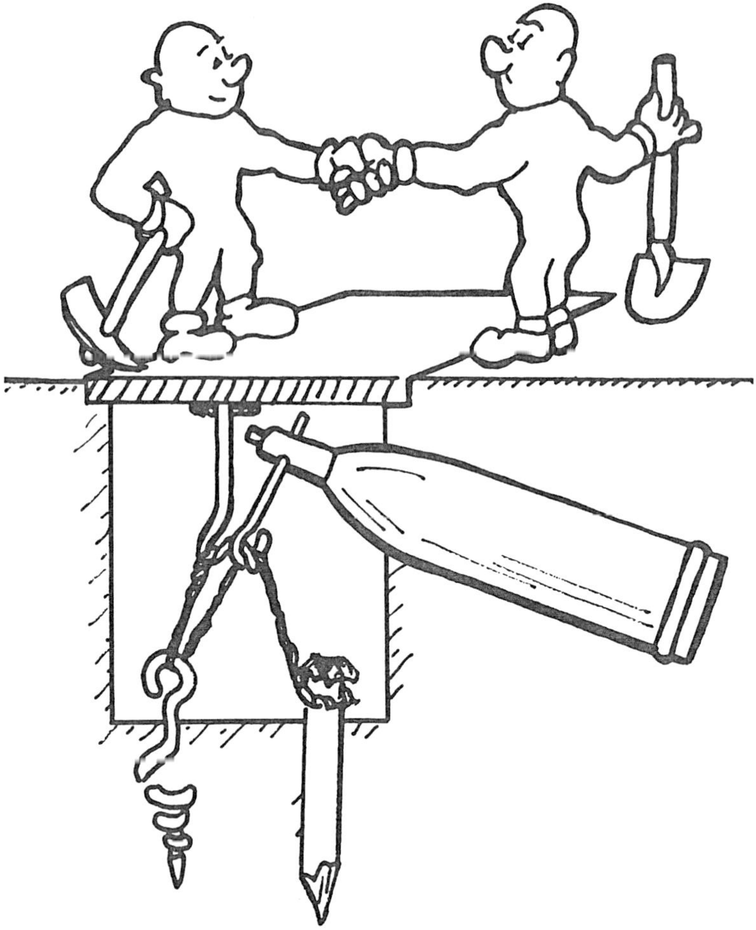

Anti-Mine-Demolition Booby Trap.

ANTI-MINE-DEMOLITION BOOBY TRAP

Type: Operates by pressure or by lifting.

Description: A cylindrical pull-type fuze is placed inside a mine or shell. A pressure plate is placed on the ground, and a small hook is attached to its underside (see sketch).

Functioning: The striker is released by a pull on the safety pin, either by pressure exerted upon the plate (the hook pulls out the safety pin), or, when used as an anti-mine-demolition booby trap: (a) If the plate is lifted, the hook releases the safety pin, but the pull cord moves and pulls the pin downward. The trap explodes. (b) In case the first trap has been discovered, and the plate and the hook have been removed, lifting the shell still causes the trap to explode by a pull on the pin by the second cord, which is attached directly to the loop in the pin.

Note: This mine may also be booby-trapped with a grenade placed underneath the base of the shell.

Neutralizing: Destroy in place.

Source: Nam-Dinh, Tonkin, 1947.

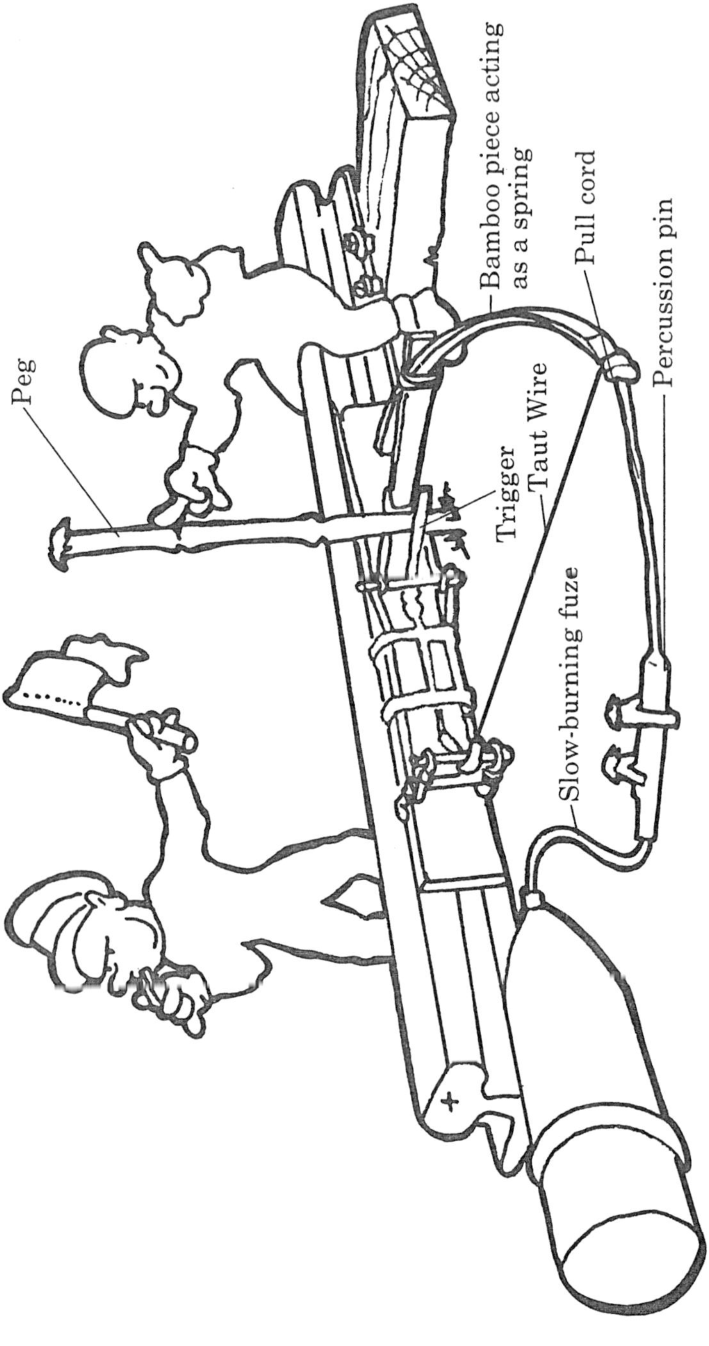

Device for Destroying a Railroad Track during Passage of a Convoy.

BOOBY TRAP FOR SABOTAGING A TRAIN

Charge: 155-mm Japanese shell.

To set: A detonator cap is placed inside the shell chamber and connected to a striker by means of a slow-burning fuze. The length of this fuze is calculated so that the charge explodes underneath the locomotive after passage of the lead cars.

The pull cord of the fuze is attached to a piece of bamboo which acts as a spring. This bamboo piece, in the shape of a semi-crossbow, is tied with a chord which passes underneath a stirrup and is attached to a stick of wood which acts as a trigger. This trigger is blocked by a peg sunk into the ground.

The passage of the first lead car of a convoy presses down on the peg supporting the trigger. The pull cord is freed and slips underneath the stirrup. The bamboo piece is released, causing a quick jerk on the cord of the percussion pin. Ignition takes place. The fire is transmitted by the slow-burning detonating fuze and the shell explodes.

Neutralizing: Cut the pull cord on the percussion pin. Remove the detonator cap from the shell. Loosen the bow and dismantle the system.

Caution: Watch out that your shins don't touch the device.

Source: Corps of Engineers T.F.I.N. Hanoi-Haiphong railroad, April, 1948.

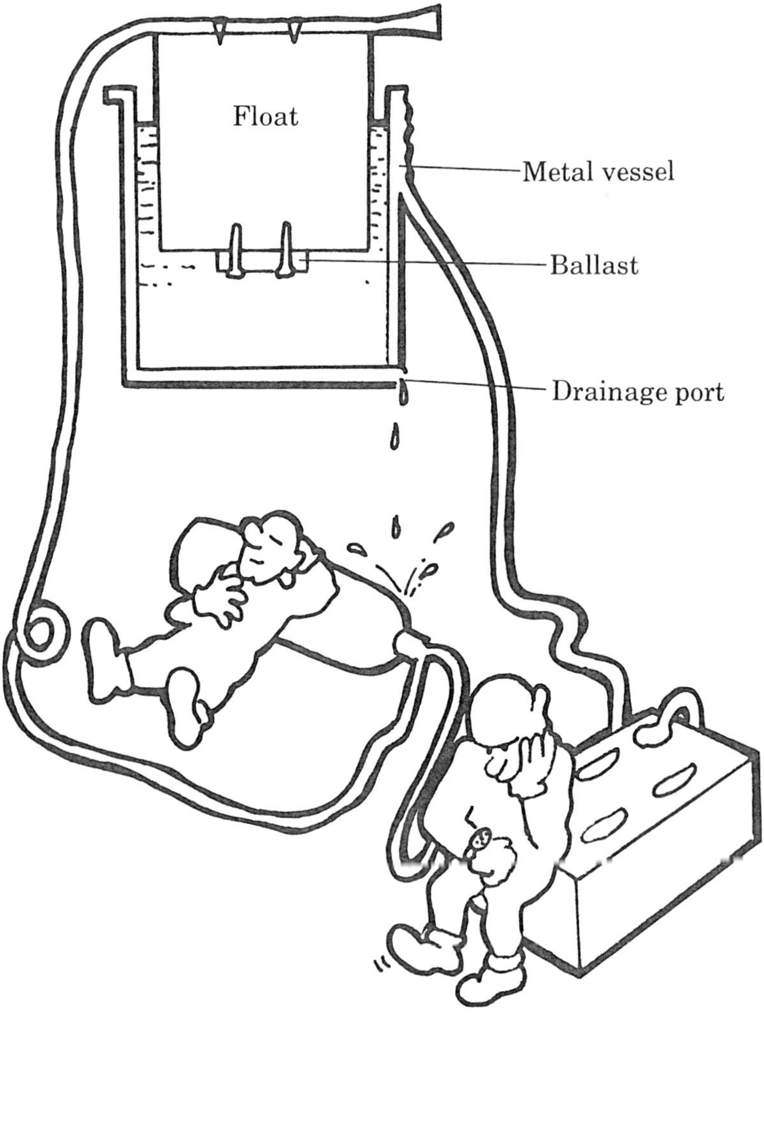

Delayed-Action Clepsydra (Water Clock) Contactor.

CLEPSYDRA (WATER CLOCK) CONTACTOR

Appearance: Any type of metal box filled with water with a float visible on the inside. An inside port allows the water to empty out.

Functioning: In the armed position, the metal vessel is filled with water. The ballasted float keeps the upper contact away from the fixed contact at the bottom of the vessel. The water empties out through the inside port.

Once the water level is lowered, the upper contact of the float touches the vessel body and the electric circuit for igniting is closed.

Note: The delay time depends on the volume of water in the vessel and the rate at which it empties.

Neutralizing: Cut the wire leading to the vessel.

Caution: Any movement of the cord on the float can cause the float to vibrate and accidentally close the electric circuit.

Source: Tan An Artillery Sector, Cochin-China, 1948.

Pit

WORDS OF ADVICE

Do not neglect any opportunity to know the enemy's weapons and, when possible, try to understand how they work.

When you discover a mine, don't call for your co-team workers to "come and see what this is." Tell them to go away and, if you can, disassemble or destroy the mine on the spot.

If you find something new, beware. If you are sure that you can disassemble it without any risk, do so. Afterwards, forward the device or a good sketch of it to the closest engineering unit so that everyone can profit from your discovery.

If you encounter a type of mine that you're not familiar with and that does not appear to be simply constructed, call the engineers.

Remember that the enemy generally prefers to destroy a road rather than repair it. Therefore, if you see some repair work or a newly filled hole where the engineers have not been working, beware.

Always avoid stopping your vehicle directly over a patch in the road. You may not always see that one of your wheels has stopped directly on a trip wire.

Beware of a pile of stones, as these greatly increase the effectiveness of a mine. The enemy knows this and very frequently uses this type of arrangement.

If you intend to destroy disarmed, but unprimed, weapons, always destroy them one at a time and not by the dozen. The best thing to do is drown the weapons in very deep water.

Remember that a mine can be caused to misfire by accidental blocking of the firing pin (e.g., by dust or rust, by an incorrectly pulled safety pin causing it to lock, etc.).

The slightest movement can cause a catastrophe. There is less danger in fooling around with a snake than there is in pulling out a safety pin any old way. Therefore, it is best to destroy a mine on the spot without touching it.

A mine that is accidentally moved (by a vehicle, a gust of wind, etc.) must be thought of as about to explode and treated as such.

Don't be tempted to start a collection of your finds.

Remember that explosives that are used as "little pranks" generally end tragically.

To avoid being sorry later, get rid of everything that is not absolutely inert.

Trip wire
Mine
Pit with booby trap

APPENDIX

RECENT TYPES OF MINES AND WEAPONS USED BY THE V. C. TROOPS

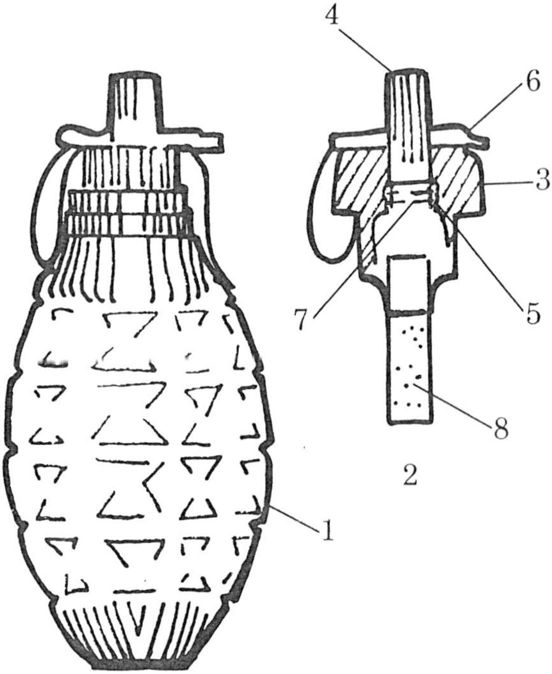
4
6
3
7
5
8
2
1

GRENADE-TYPE MINE

Type: Antipersonnel mine.

Description:

Body:

(1) cast iron with serrations for fragmentation effect (shaped like a defensive hand grenade).

Charge: Either black powder or an explosive.

Fuze:

(2) instantaneous, mechanical pressure-type, consists of:
(3) steel body;
(4) striker;
(5) safety spring;
(6) safety pin;
(7) primer;
(8) detonator.

Total weight: 0.8 kg
Weight of explosive: 0.1 kg
Diameter: 6 cm
Height: 13 cm
Functioning: Once the safety pin is removed, pressure on the striker ignites the primer.

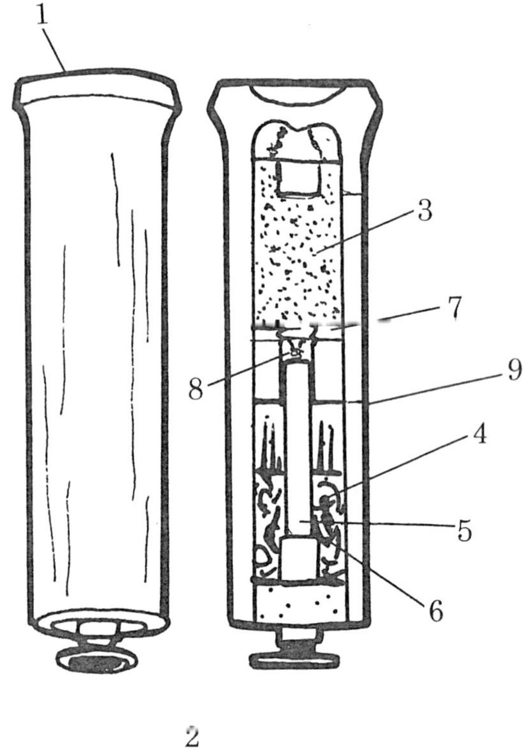
1
3
7
8
9
4
5
6
2

BAMBOO MINE

Type: Antipersonnel mine.

Description:

Body:

(1) consists of a piece of bamboo stick with
(2) wooden obturators.

Charge:

(3) small roll of compressed black powder.

Projectiles:

(4) makeshift projectiles consisting of stones and pieces of glass or porcelain.

Fuze: Instantaneous pressure-type fuze consisting of:

(5) wooden striker with metal tip;
(6) guide tube for wooden striker;
(7) separator section made of wood, consisting of:
(8) primer;
(9) hole for trip wire.

Total weight: 0.5 kg (approx.)
Weight of explosive: 0.1 kg (approx.)
Diameter: 5 to 6 cm (approx.)
Height: 22 cm (approx.)
Functioning: The mine is buried in a vertical position with the striker in position. Pressure on the striker causes the primer to break and the mine to explode. If a pile of material is placed around the mine, the projectiles will shoot upwards.

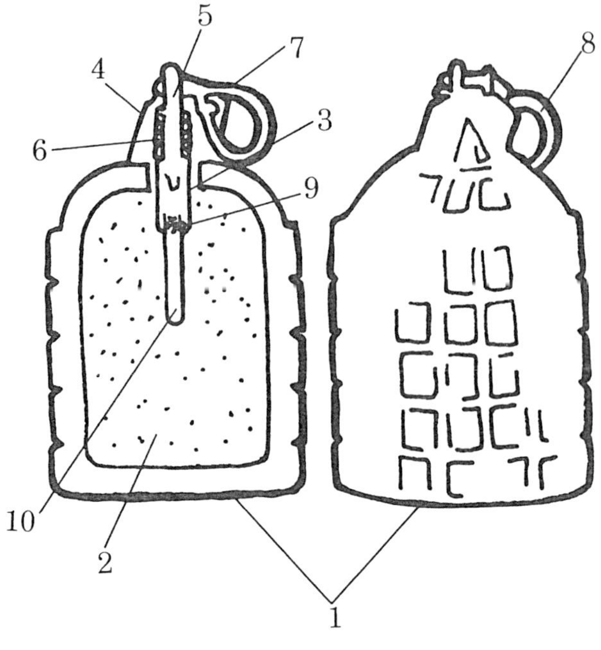
5
7
8
4
3
6
9
10
2
1

WIDE-MOUTHED BOTTLE MINE

Type: Antipersonnel and antivehicular mine.
Description:

Body:

(1) cast iron with serrations for fragmentation effect. Shaped like a wide-mouthed bottle.

Charge:

(2) ammonium nitrate explosive.

Fuze:

(3) instantaneous mechanical fuze with pull-type safety pin. The fuze is made up of:
(4) light-alloy body;
(5) striker;
(6) striker spring;
(7) striker lever;
(8) starter pin;
(9) primer;
(10) detonator.

Total weight: 3.8 kg
Weight of explosive: 0.85 kg
Diameter: 11 cm
Height: 17 cm
Functioning: A pull on the trip wire causes the safety pin to come out. The lever releases the striker which fires the primer.

Observations: The explosive used in this mine is a very shattering type that is little known and little used. It has some similarities to the ammonals.

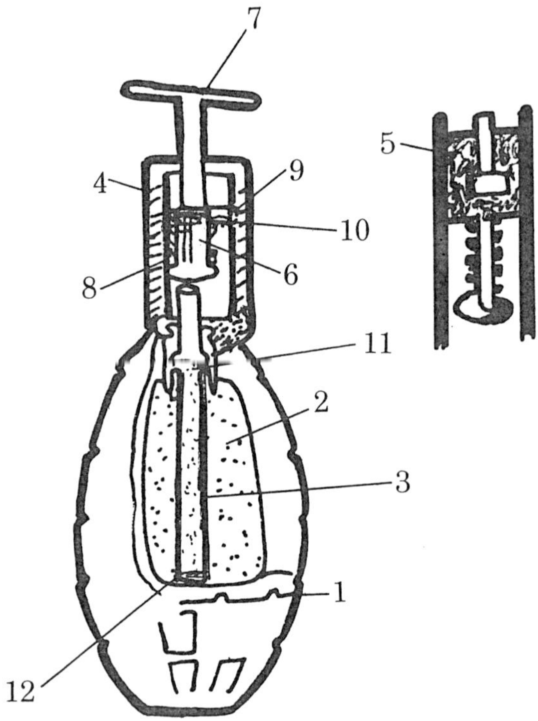
7
5
4
9
10
6
8
11
2
3
1
12

PINEAPPLE MINE

Type: Antipersonnel mine.

Description:

Body:

(1) cast iron with serrations for fragmentation effect. Shaped like a pineapple.

Charge:

(2) ammonium nitrate explosive;
(3) powdered melinite relay.

Fuze: instantaneous mechanical pull-release fuze, consisting of:

(4) sheet-metal frame;
(5) plug for the cap and striker chambers;
(6) striker;
(7) T-shaped hammer with notch in the handle for a bolt;
(8) hammer spring;
(9) bolt;
(10) bolt spring;
(11) primer;
(12) detonator.

Total weight: 2 kg
Weight of explosive: 0.27 kg
Diameter: 9 cm
Height: without fuze: 12 cm
with fuze: 21 cm

Functioning:

Pull-type: In the armed position, the hammer spring is compressed and the bolt is fitted into the notch on the handle. A pull on the pull cord, which is tied to one of the arms of the hammer, causes the hammer to pivot. The bolt is released and the primer is ignited.

Release-type: In the armed position, the hammer spring is held tight by a pull cord without using the bolt. Breaking this cord causes the mine to function.

Observations: The striker, which is removable, is placed in position only when the fuze is armed.

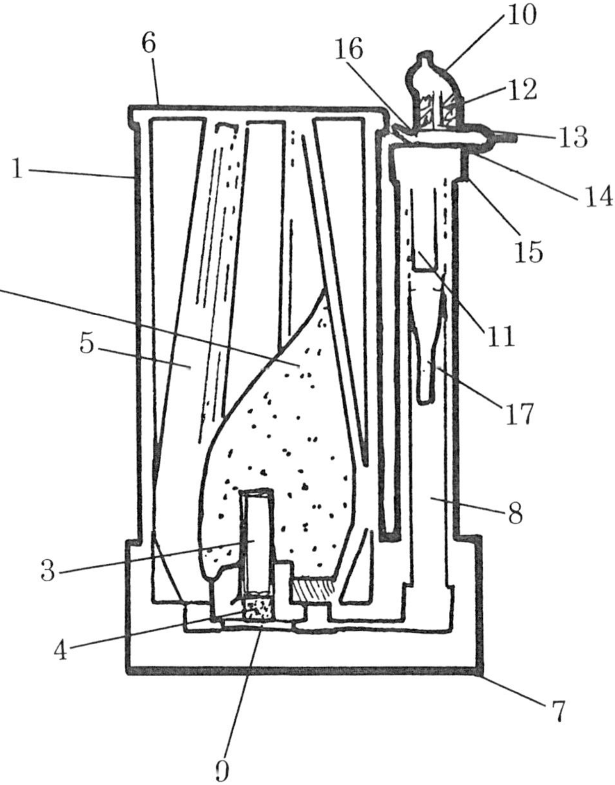
10
6
16
12
13
1
14
15
2
11
5
17
8
3
4
7
9

BOUNDING MINE

Type: Bounding antipersonnel mine.

Description:

Projectile, consisting of:

(1) 81-mm explosive mortar projectile without fins;
(2) explosive charge;
(3) detonator;
(4) black powder delay.

Cannon tube, consisting of:

(5) metal launch tube;
(6) metal cover;
(7) metal base;
(8) cap chamber tube.

Spill charge:

(9) smokeless powder.

Pull-type instantaneous mechanical igniter consisting of:

(10) body;
(11) striker;
(12) striker spring;
(13) release;
(14) safety bolt;
(15) safety grip;
(16) safety pin;
(17) primer cartridge.

Total weight: 7.35 kg
Weight of projectile: 3.1 kg
Weight of cannon tube: 5.1 kg
Weight of fuze: 0.15 kg
Weight of spill charge: 0.009 kg
Height: 22 cm
Diameter: 14 cm
Functioning: Once the safety pin and safety grip are removed, a pull on the pull cord causes the release to pivot and the bolt to come out. The striker is freed and strikes the primer cartridge. The spill powder drives the projectile and ignites the delayed-action fuze. The projectile explodes above ground.

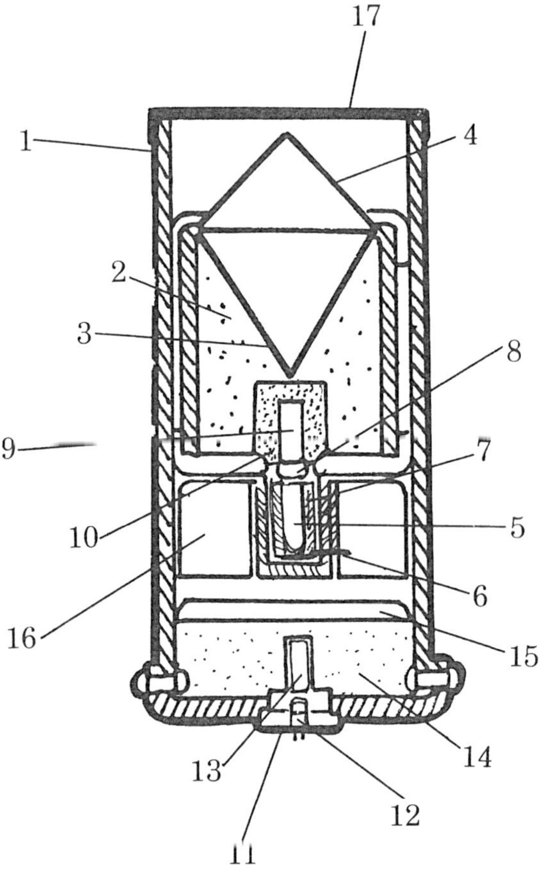
17
1
4
2
3
8
9
7
10
5
16
6
15
13
14
12
11

BOUNDING BAZOO MINE

Type: Bounding antivehicular mine with hollow charge.

Description:

Bazoo mine consisting of:

(1) steel cylinder body;
(2) melted melinite explosive charge;
(3) cone-shaped bronze casing for hollow charge;
(4) sheet-metal cover;
(5) striker;
(6) safety pin;
(7) safety spring;
(8) primer:
(9) detonator;
(10) powdered melinite primer relay.

Cannon tube (11), containing:

(12) electric fuze;
(13) igniter powder;
(14) black powder spill explosive;
(15) cardboard partition;
(16) separator;
(17) metal lid.

Total weight: — —
Weight of explosive: 1 kg
Weight of spill charge: — —
Diameter of mine: 15 cm
Height of mine: 25 cm
Diameter of cannon tube: 17 cm
Height of cannon tube: 38 cm
Functioning: The bazoo mine is placed in the cannon tube without the safety pin. Setting off the electric fuze causes the spill charge to ignite, which thrusts the bazoo mine out of the cannon tube. As soon as the bazoo mine hits an object, the striker functions by inertia.

Observations: There is also a nonleaping bazoo mine in which the mechanical striker-type fuze is replaced by an electric fuze.

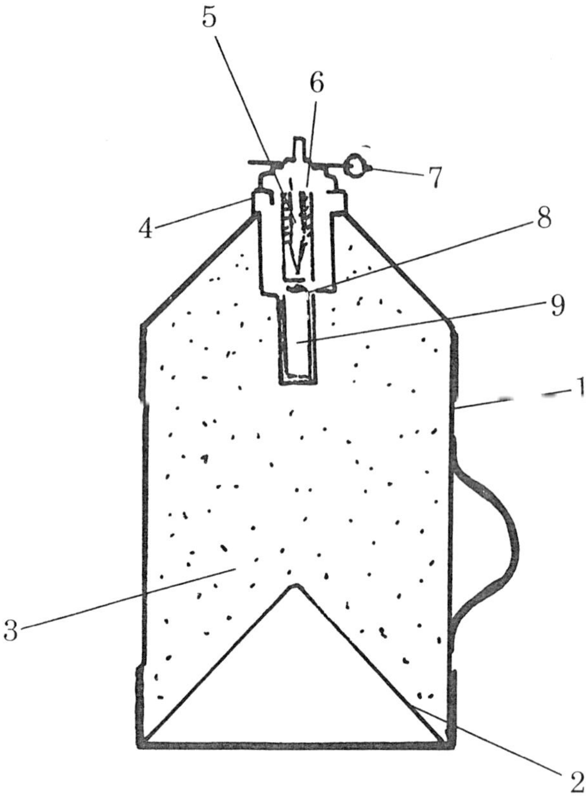
5
6
7
4
8
9
1
3
2

MINE WITH HOLLOW CHARGE

Type: Antivehicular and antipersonnel mine.

Description:

Mine:

(1) cylindrical sheet-metal body;
(2) metal lining of cone-shaped hollow charge.

Charge:

(3) nitrate explosive.

Fuze:

(4) steel body;
(5) striker;
(6) striker spring;
(7) starter pin;
(8) primer;
(9) detonator.

Total weight: 5 kg (approx.)
Weight of explosive: 3.5 kg (approx.)
Diameter: 12 cm
Height: 21 cm

Functioning: A pull on the pull cord pulls out the starter pin. The striker is released and strikes the primer.

Observations: There is an identical demolition mine equipped with folding legs that enable it to be placed at the optimum yield distance from a wall to be destroyed.

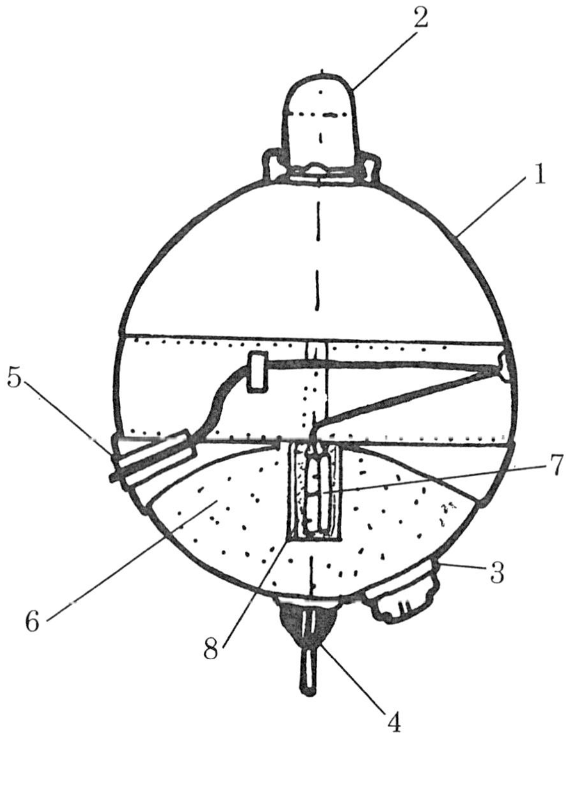
2
1
5
7
3
6
8
4

UNDERWATER MINE I

Type: Mine for use against river boats

Description:

Mine body:

(1) spherical body made of 4-mm sheet metal;
(2) carriage ring;
(3) attachment ring;
(4) sealer plug;
(5) stopper plug for inserting conductor wires.

Explosive charge:

(6) melted melinite.

Detonator:

(7) two electric primers with different resistances;
(8) initial priming relay (powdered melinite).

Total weight: 43 kg
Weight empty: 19 kg
Weight of explosive: 24 kg
Diameter: 0.52 m

Functioning: Remotely detonated. The electric current causes the primers to ignite and the mine to explode.

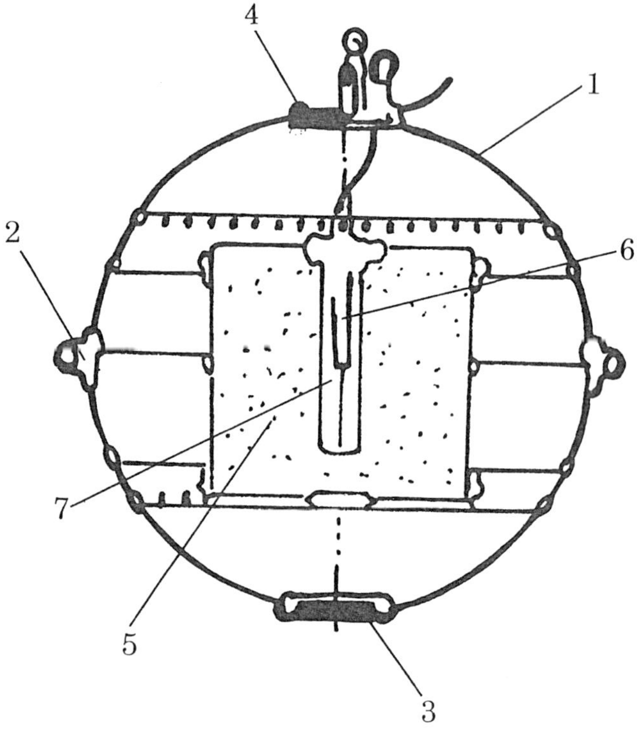
4
1
2
6
7
5
3

UNDERWATER MINE II

Type: Mine for use against river boats

Description:

Mine body:

(1) spherical body made of 2-mm-thick sheet metal;
(2) attachment rings;
(3) sealer plug;
(4) stopper plug for inserting conductor wires.

Charge:

(5) ammonium perchlorate.

Detonator:

(6) electric primer;
(7) powdered melinite primer relay.

Total weight: 42 kg
Weight empty: 20 kg
Weight of explosive: 22 kg
Diameter: 0.63 m
Height: 0.58 m
Functioning: Remotely detonated. The electric current causes the primer to ignite and the mine to explode.

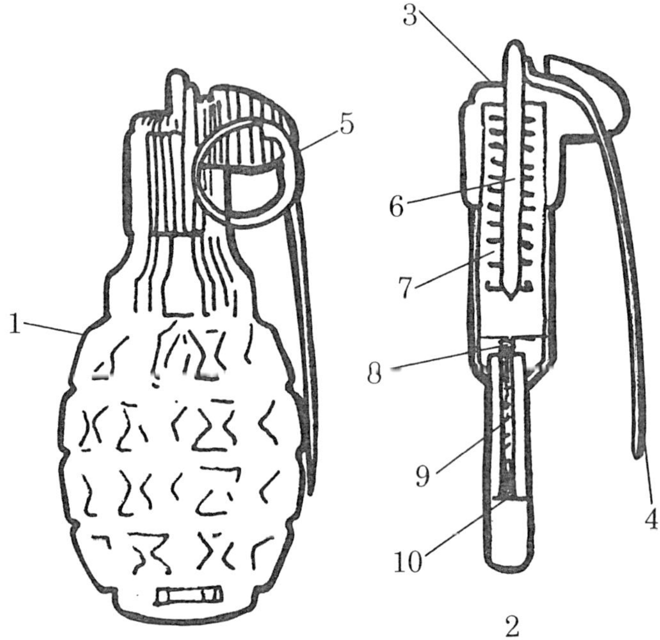
1
5
3
6
7
8
9
10
4
2

TIME-FUZE HAND GRENADE I

Type: Antipersonnel defensive grenade

Description:

Grenade body:

(1) cast iron with serrations for fragmentation effect.

Charge: black powder.

Detonation:

(2) time fuze-type fuze plug, consisting of:
(3) light alloy body;
(4) release lever;
(5) safety pin;
(6) striker;
(7) striker spring;
(8) primer;
(9) slow-burning fuze;
(10) black powder relay.

Total weight: 0.66 kg
Weight of body: 0.5 kg
Weight of fuze plug: 0.1 kg
Weight of explosive: 60 gr
Diameter: 60 mm
Height: 125 mm
Time delay: about 5 seconds
Functioning: Removing the safety pin frees the release lever and allows the striker to fire the primer. The delayed action fuze burns for about 5 seconds.

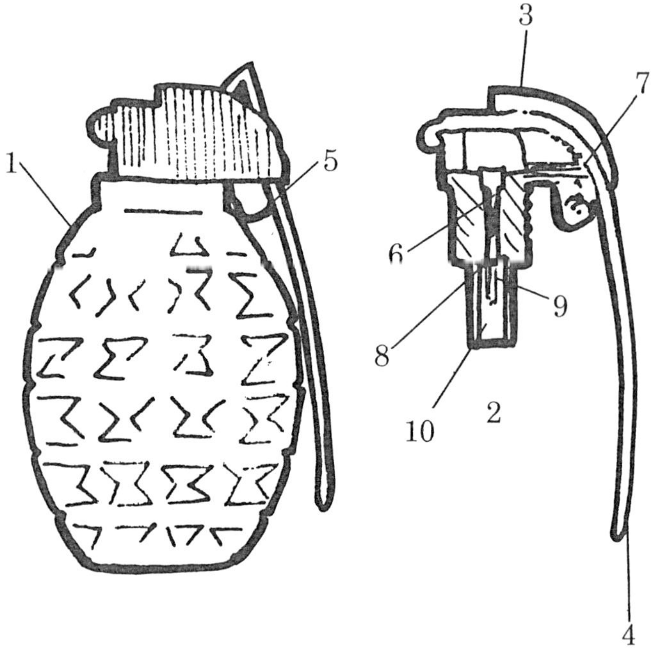
1
5
3
7
6
9
8
10
2
4

TIME-FUZE HAND GRENADE II

Type: Antipersonnel defensive hand grenade.

Description:

Grenade body:

(1) cast iron with serrations for fragmentation effect.

Explosive charge: black powder.

Detonation:

(2) time-fuze-type fuze plug, consisting of:
(3) light alloy body;
(4) release lever;
(5) safety pin;
(6) striker;
(7) striker pin;
(8) primer;
(9) slow-burning fuze;
(10) detonator.

Total weight: 0.65 kg (approx.)
Weight of body: 0.49 kg (approx.)
Weight of fuze plug: 0.1 kg (approx.)
Weight of explosive: 60 gr (approx.)
Diameter: 55 mm
Height: 100 mm
Time delay: about 5 seconds
Functioning: The safety pin is removed, the release lever is freed, and the striker fires the primer. The delayed action fuze burns for about 5 seconds.

Observations: The fuze plug is a copy of an American fuze.

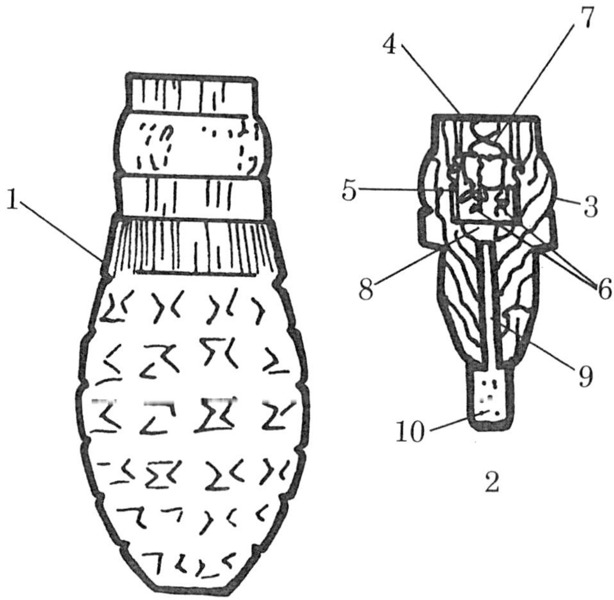
1
4
7
5
3
8
6
9
10
2

PULL-TYPE HAND GRENADE

Type: Antipersonnel defensive grenade.

Description:

Grenade body:

(1) cast iron with serrations for fragmentation effect.

Explosive charge: black powder.

Detonation:

(2) pull-type time-fuze plug, consisting of:
(3) wooden piece attached to the body by two pins;
(4) metal cover;
(5) capsules containing the fuzing material;
(6) percussion pins;
(7) cord and pull ring;
(8) igniter paste;
(9) slow-burning fuze;
(10) black powder relay.

Total weight: 0.75 kg
Weight of body: 0.53 kg (approx.)
Weight of fuze plug: 0.15 kg (approx.)
Weight of explosive: 70 gr
Diameter: 55 mm
Height: 115 mm
Time delay: about 7 seconds
Functioning: Remove the metal cap, pull on the pull ring, and toss the grenade. The friction of the percussion pins causes the fuzing material in the capsules to ignite. The igniter paste is lit and in turn ignites the slow-burning fuze.

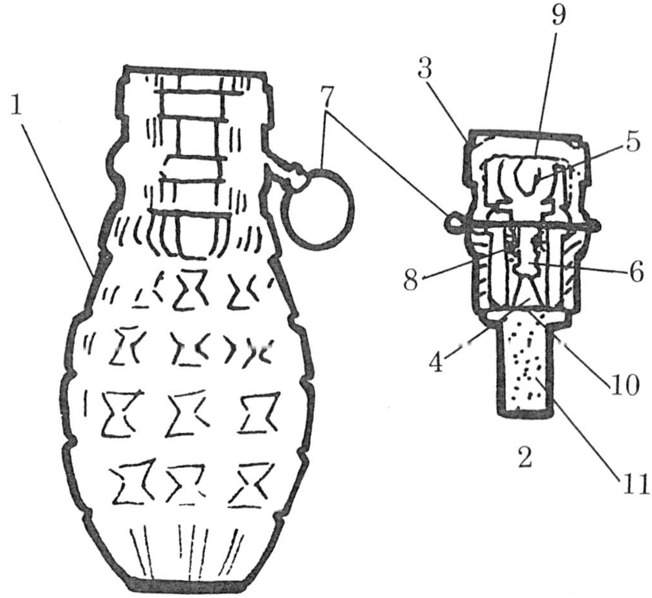
1
7
3
9
5
8
6
4
10
2
11

PERCUSSION HAND GRENADE

Type: Antipersonnel defensive grenade.

Description:

Body:

(1) cast iron with serrations for fragmentation effect.

Explosive charge: black powder.

Detonation:

(2) percussion-type fuze plug, consisting of:
(3) tin body;
(4) cone-shaped inner stop plug;
(5) lead ball;
(6) striker with hollow head;
(7) safety pin;
(8) safety spring;
(9) movable cap chamber;
(10) primer;
(11) black powder relay.

Total weight: 0.645 kg
Weight of body: 0.415 kg
Weight of fuze plug: 0.17 kg
Weight of explosive: 60 gr
Diameter: 60 mm
Height: 115 mm
Functioning: With the safety pin pulled, the fuze plug functions at the least shock and at any angle of fall (heavy ball, movable cap chamber and conical notches in the stop plug, and the striker head).

Observations: This fuze plug is a copy of the British "Allewais" fuze plug used to ignite " Gammon bomb" grenades. Note that throwing this grenade presents a certain amount of danger to the thrower, as the safety spring is very weak and there is no safety gasket.

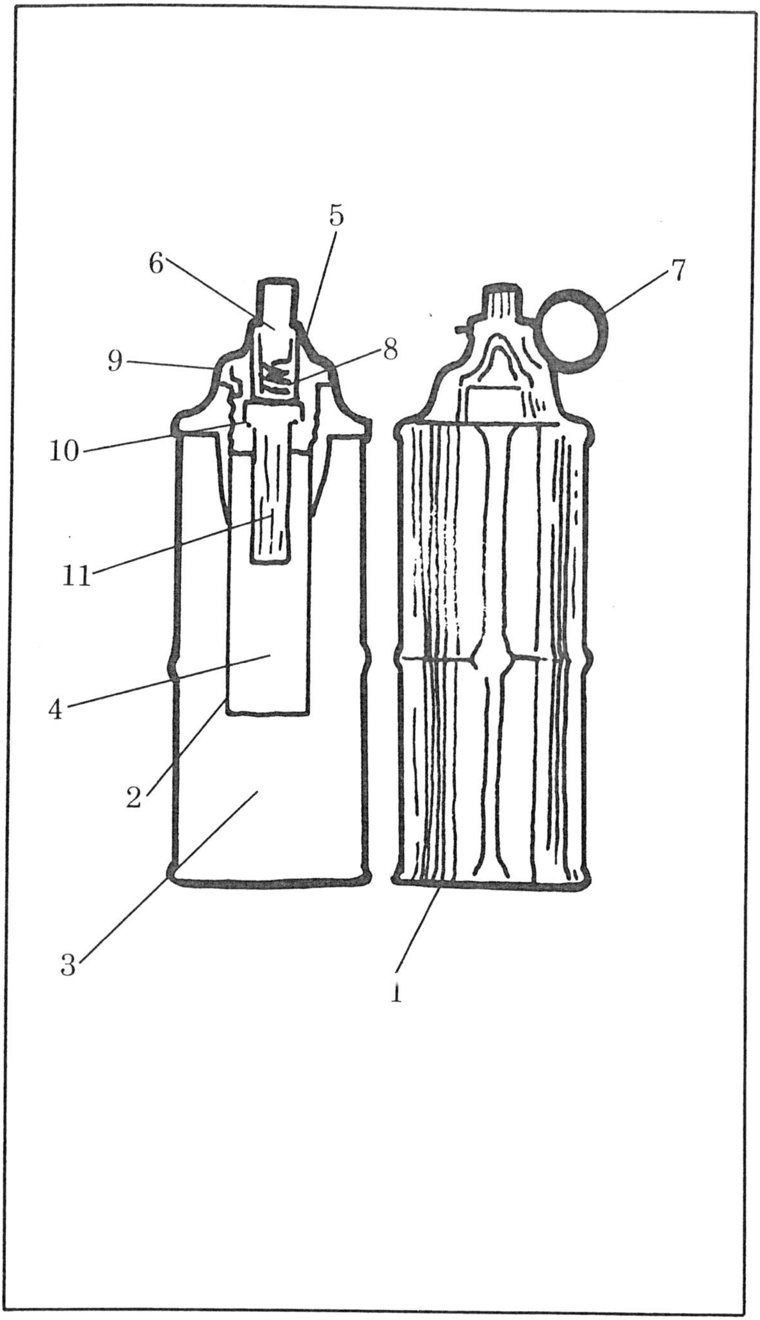
5
6
7
9
8
10
11
4
2
3
1

INCENDIARY HAND GRENADE

Type: Incendiary grenade.

Description:

Grenade body:

(1) cylindrical sheet-metal body;
(2) relay casing filled with black powder.

Charge:

(3) incendiary composition (solution of latex in benzine);
(4) igniter charge (black powder).

Detonation (time-fuze-type fuze plug):

(5) fuze body;
(6) movable striker;
(7) safety pin;
(8) safety spring;
(9) vent;
(10) fulminating cap;
(11) slow-burning fuze.

Total weight: 0.8 kg (approx.)
Weight of igniter charge: 20 gr
Weight of incendiary composition: — —
Time delay: 4 seconds
Diameter: 53 mm
Height: 170 mm
Functioning: Remove the safety pin. Hit the striker against a hard object. Toss the grenade. The slow-burning fuze burns for about 4 seconds and detonates the black-powder igniter charge. The incendiary composition is lit and scattered.

Observation: Can also be considered an explosive-type grenade.